Matthew Warren is an economist and journalist. He has spent the past 15 years working for the electricity, downstream gas, renewable energy and coal industries. He was Chief Executive of the Australian Energy Council, the Energy Supply Association of Australia and the Clean Energy Council. He was also Environment Writer at *The Australian* newspaper and worked for the New South Wales Minerals Council. He likes trail running and tabouli. This is his first book.

BLACKOUT

books that leave an impression

Published by Affirm Press in 2019
28 Thistlethwaite Street, South Melbourne, VIC 3205
www.affirmpress.com.au
10 9 8 7 6 5 4 3 2 1

Text and copyright © Matthew Warren, 2019
All rights reserved. No part of this publication may be reproduced without prior permission of the publisher.

Title: Blackout / Matthew Warren, author.
ISBN: 9781925870176 (paperback)

 A catalogue record for this book is available from the National Library of Australia

Cover design by Design by Committee
Typeset in Minion Pro 12 / 17.5 by J&M Typesetting
Proudly printed in Australia by Griffin Press

The paper this book is printed on is certified against the Forest Stewardship Council® Standards. Griffin Press holds FSC chain of custody certification SGS-COC-005088. FSC promotes environmentally responsible, socially beneficial and economically viable management of the world's forests.

BLACK OUT

HOW IS ENERGY-RICH AUSTRALIA RUNNING OUT OF ELECTRICITY?

MATTHEW WARREN

Contents

Foreword   *ix*
Preface   *xiii*

Introduction   1
1. Where does electricity come from?   19
2. How did climate change affect electricity in Australia?   43
3. What caused the electricity crisis?   67
4. How does the electricity grid work?   89
5. What are renewables?   115
6. What is firm power?   139
7. Why is there so much rooftop solar in Australia?   165
8. Can renewables grids ever be reliable?   181
9. What can consumers do?   201
10. Where do we go from here?   229

Acknowledgements   *257*
Selected references   *259*
Index   *267*

Foreword

Electricity is conceptually so simple. Electrons drift serenely in wires like water molecules flow smoothly in pipes. But by the time we integrate an electricity system into a modern economy, horrendous complexity emerges. The system operators and regulators must juggle the operation of competitive markets so that prices are minimised, supply is guaranteed and the laws of physics are obeyed. If the juggler is hesitant for an instant, the laws of physics prevail and the system crashes.

In *Blackout*, Matthew Warren brings to bear his experience and narrative skills to explain this complexity to novices and experts alike. The interaction of politics, physics, psychology, markets and economics is laid bare, and from the detritus of battles fought between reactionaries in the red corner and ideologues in the green corner a way forward is forged.

– Australia's Chief Scientist Dr Alan Finkel

For Harriet and Tully

Preface

Blackout is the story of the electricity crisis in Australia. It's the synthesis of my professional life over the past 15 years, which, quite by accident, I have spent working on the impact of climate change and the future of energy in Australia. Over this time, I worked at the NSW Minerals Council as a lobbyist for the (mainly coal) mining industry, then as environment writer for *The Australian* newspaper. After that I ran the Clean Energy Council lobbying for the renewables industry, then the Energy Supply Association where I represented the electricity industry and downstream gas businesses. Finally I was in charge of the Australian Energy Council, which represents electricity generators and retailers. In some ways it's an unusual CV.

I've spent most of this time first trying to understand and then to explain how electricity works in Australia, what the impact of climate change will be on this crucial industry, and what we need to do about it. Over this time my views and personal opinions have evolved, sometimes significantly. Along the way I've tried to avoid being captured by any ideology. About the closest I come to any belief system is that I do think well-designed markets are efficient ways of solving most problems, if allowed to work.

This book is basically a translation of the complex and, at times, unfathomable technical language of electricity policy and engineering into plain English. I'm an economist and journalist by training, so at first I didn't speak Electricity either. I had to learn it on the job. I'm still only partially fluent, and still need help translating some of the more complex systemic issues. I'm indebted to the engineers, scientists and other technical experts who have shared their knowledge with me over the years and hope my translation is sufficiently accurate as to be informative and not misleading.

For years, Australia has been gripped by a seemingly interminable argument over what to do about climate change, which unavoidably means what to do about electricity. Both these issues are complex, and both climate science and electricity grids sit outside the conventional 20th-century left-versus-right construct. The warnings on climate change are simply the consensus professional advice from a school of scientists. The science itself is apolitical. It's the interpretation and leverage of the scientific warning that has politicised climate change since it entered mainstream consciousness in the mid-2000s.

What I have observed, as the public debate has rolled along for more than a decade, is how inaccessible the discussion around climate and electricity has been for a lay audience. The political debate is played out regularly for us, but the technical explanations behind it don't fit in conventional media. Most journalists, even those regularly covering it, can only afford the time to develop a cursory understanding. Outside a handful of core ministers and their staff, politicians are similarly bewildered.

Energy was a largely bipartisan issue until the mid-2000s, and recognition of climate science has been bipartisan in most developed economies. In lieu of understanding, we have transformed machines into ideologies. Coal is held up as a metaphor for common sense, with talk of climate change and use of renewables represented as nothing more than political correctness gone mad. Belief in renewables is similarly unconditional. Renewables have become a symbol of change, the relatively obvious gaps in their performance waved away with a utopian view of the world, where problem-solving technologies simply arrive cheaply and on time, like Uber Eats. The two sides bicker on Twitter all day long about which of them caused the latest price spike or blackout. This all makes about as much sense as the great Ford-versus-Holden rivalry of the late 20th century.

If we are going to keep having a public debate about climate change and electricity in Australia, then it might help to be better informed. The story of *Blackout* is hopefully useful to those who want to get their head around how we got into this mess and how we can get out of it. The information contained herein comes from a selection of the hundreds of technical reports produced on these subjects, coupled with the knowledge gleaned from my various front-row seats as the debates unfolded.

By way of examples, I was writing at *The Australian* in late 2006 when climate change exploded on the then prime minister John Howard in the form of drought-ravaged suburban Australian gardens. I was at the climate talks in Bali in 2007 when Kevin Rudd ratified the Kyoto Protocol (it looked much more impressive than it really was). I later lobbied for the implementation of the Renewable Energy Target (RET) and watched as the renewables industry was itself stunned by the runaway success of rooftop solar panels. Later, for the electricity industry, I worked with major companies aligned with environment groups and others to push for meaningful climate policy – not out of any environmental sensibilities, but because they needed it in order to run their businesses. With each failed attempt, the industry's frustration grew.

Well before 2016, when South Australia suffered a major blackout, many in the broader electricity industry were becoming increasingly concerned at the unplanned growth of large-scale renewables generation in a small, relatively isolated grid like South Australia's. On the day of the blackout, I was briefing international industry counterparts on the living experiment in South Australia and speculating on what might go wrong. Later that night, Lisbon time, it did.

I don't think this needs to be as hard as we are making it. Energy in Australia in the 21st century is an important issue. If we could move past the ideology and implement the required technical reforms, the real debate is about what the Australian economy looks like as our energy systems change. The sooner we get past the politics and get on with the real stuff, the better.

INTRODUCTION

Electricity is modernity. Electricity is ubiquitous. It's like a magic spell that surrounds us. It lights our homes, streets and workplaces. Our leisure is electrified: sporting venues, cinemas, theatres, bars and restaurants. Computers have reshaped our lives and how we work, communicate and learn; they run on electricity. We learn in electrified classrooms. We sleep on electric blankets. Often the first and the last thing we do every day is turn electrical devices off or on. Electricity is *so* ubiquitous that it has become invisible. We know it's there, but we don't even think about it – at least until it's not there. We don't consume electricity; we consume the devices it powers. Electricity is the great enabler and the ordinary person's butler. Always in the background, it quietly attends to the many and various needs of our busy, modern lives. No matter how much we require, it is always willing to serve.

The term 'modernism' refers to a cultural and philosophical revolution in the western world that coincided with the electrification of these societies. When we think of something being 'modern', intuitively it runs on, or uses, electricity. Electricity networks were first rolled out in the cities and towns of most developed countries, including Australia, towards the end of the 19th century and this continued through the 20th century. Initially electricity was used primarily for lighting, then refrigeration, then cooking, transport, cleaning, heating and cooling, radio, then televisions, computers and smartphones. The more than 900 million people on Earth estimated by the International Energy Agency to still not have electricity do not live modern lives: they live in poverty.

Modern life requires abundant and uninterrupted electricity. Electricity enables increasingly clever devices and machines. It gives us effortless control over our lives. For more than a century, the citizens of entire cities, towns and countries have been able to work and play from daylight into the darkness, listen to the radio and watch television, stay warm when it's cold or cool when it's hot, and walk home more safely in lit streets. Modern people have saved hundreds of hours each year by using electrical machines to do menial tasks such as washing clothes and dishes. Electricity has enabled ordinary people to live richer lives in most respects than the grandest emperors and queens of human history. We don't even think about how remarkable this is. We just think this is normal.

At 3.50 on a Wednesday afternoon in September 2016, the entire state of South Australia went black. No houses or buildings had power. All the traffic lights went out. Hospitals kept patients alive using emergency generators. Hotels couldn't serve beer, fridges stopped being cold, sliding doors couldn't open, lifts got stuck and nothing worked for more than three hours across an entire state. This was not a drill: 1.7 million modern people were suddenly without power. Most of the city of Adelaide was brought back on later that night, but parts of the state were without electricity for days.

Blackouts are supposed to be a quaint and charming feature of travelling in developing countries, or a childhood memory from the Australia of the 1960s and '70s. Electricity systems can, very occasionally, fail, but such failures should be isolated and exceptional. Suddenly that seems to have changed in Australia. Electricity prices have doubled over the past decade. Since South Australia's system black, people have been talking about whether there is enough power to get through the next summer heat wave. The electricity systems in entire regions, such as South Australia, appear to be failing.

How can a service that has been provided for a century suddenly start failing now? How can it get so expensive when other comparable services, such as telephone calls and internet data, are getting cheaper? In 2006, John Howard described Australia as a potential 'energy superpower': the biggest exporter of coal in the world, soon to be the biggest exporter of gas, and with abundant wind and sunshine to power renewable technologies. Australia also has the world's largest uranium reserves. So, given all of this, what has gone wrong? Why are we suddenly talking about electricity as a problem?

This is the story of the electricity crisis in Australia. Like most big stuff-ups, this crisis wasn't the result of any single decision or event. A series of political decisions, all imposing technical challenges on the electricity system, combined and were then amplified by Australia's unique geography. The speed at which things have tumbled out of control is a warning to the rest of the world. This could have been avoided, and can still be defused, but it will require cool heads, genuine political leadership, and a return to relying on genuine experts and dedicated institutions to run the grid efficiently. If allowed to, they will light the pathway to a safe exit from this crisis.

The largest electricity archipelago in the world

Australia is the world's biggest island. It's also the world's biggest electricity archipelago, made up of more than 1000 separate electricity systems: from remote cattle stations and the resort at Uluru, to a single grid covering the entire east coast. All of these grids are electricity islands. They operate completely independently of each other.

The biggest of these electricity islands, and the ground zero of the current electricity crisis, is what is officially called the National Electricity Market (NEM). Around 21 million Australians live and work inside the NEM, a gigantic machine that generates and moves electricity around the eastern seaboard of Australia, from Ceduna on the coast of South Australia to Port Douglas in Far North Queensland, some 2500 kilometres away. Through a series of undersea cables and overhead lines, it reaches down to Bruny Island off the coast of Tasmania, into Broken Hill in western New South Wales and up to Barcaldine in Central Queensland.

This single 'national' grid was created by connecting the five eastern states' electricity islands. Each state grid was originally developed during the 20th century by private businesses, then governments, spreading outwards like a spider web from the major cities and towns. The islands were united only recently – the first electricity bridge was the Snowy Mountains hydroelectric scheme, which linked New South Wales and Victoria. Then a cable linked South Australia and Victoria in 1989. Joined together, they made what resembles a huge coathanger of poles and wires crisscrossing the eastern edge of this vast continent. Put simply, it is the largest electricity grid in the world.

Its little sister, the South West Interconnected System (SWIS), delivers electricity to the Greater Perth region. From there the islands in the archipelago get smaller and smaller: Darwin and Alice

Springs have their own grids, as do Broome, Carnarvon, Coober Pedy and Birdsville. There are micro-grids on major populated islands such as King and Flinders Islands in Bass Strait, and in hundreds of remote locations, homesteads and communities.

All these grids, large and small, were installed, expanded and refined during the 20th century. Blackouts were common in the early days of electrification, but by the end of the millennium, the system in Australia was humming like a Swiss watch. In the year 2000, the typical household electricity bill was stable at around $800 to $900 a year. Electricity reliability was in line with most other developed countries, meeting the target of 99.998 per cent (an acceptable outage level of 0.002 per cent of total supply). Most outages that did occur were small and localised: usually a fault somewhere in the network of poles and wires that was quickly repaired.

By then, the state governments of Victoria and South Australia had begun to sell off networks and power stations to private companies. The eastern states were linked by huge transmission lines and opened up to competition via a market-based system of electricity trading. All this happened without noticeable public fanfare: in other words, you probably don't remember these changes or didn't hear about them. Reliability and prices didn't materially change (except for a lot of businesses, for which the cost of electricity fell significantly). At the time there was considerable political debate about the privatisation, but the arguments were more ideological than operational. Essentially the same people ran the same systems, they just worked for private companies rather than government departments. The grid was not plunged into crisis because of privatisation and the introduction of competitive electricity markets; actually, the evidence at the time suggested that generators worked slightly better with increased focus on their performance. The grid hardly blinked.

In 2000, most of Australia's east-coast electricity system was built around a fleet of huge coal-fired power stations. These were located in coal-rich regions, conveniently adjacent to the big cities they supplied – the south-east and central coast of Queensland, the Hunter Valley in New South Wales, the Latrobe Valley in Victoria and the 'Iron Triangle' in South Australia. At their zenith, coal-fired power stations produced around 83 per cent of our total electricity. They ran, and mostly still run, twenty-four hours a day, seven days a week, dialling up to meet peaks in electricity demand that occur every morning and evening, and during summer heat waves and winter cold snaps.

Mainland east-coast electricity supply was topped up from two main sources: a patchwork of smaller gas-powered generators whose job it was to switch on during times of very high demand, and hydro generation from the Snowy Mountains. In 2000, the coal and gas used were abundant and cheap, and the hydro was provided by rainfall. It was, by international standards, about as cheap and reliable an electricity system as you could build. Its brutal simplicity, reliability and low cost had attracted global industries including aluminium and other metals processors. These were 'the good old days' of cheap and reliable electricity in Australia.

The climate problem

But there was a problem: greenhouse gas emissions. Those coal-fired power stations had been built during the postwar era, underpinning the rapid transformation of the economy. Australian electricity demand increased eightfold from 1955 to 1985. This fast growth in electricity demand tracked the equally fast growth of the Australian economy. Cheap electricity transformed not only household life, but the entire country. The abundance of vast, high-quality coal seams so proximate to our major cities was seen as a godsend. In the 1960s, '70s and '80s, when most of these massive power stations were built, it was a no-brainer. Why would you use anything else?

Towards the end of the 20th century, the jungle drums were starting to beat more loudly on greenhouse gas emissions and the threat they posed to the planet. For Australia this was particularly inconvenient: its coal-dominated electricity system was, per capita, the world's biggest emitter of these gases.

Climate change is divisive. Essentially it is a scientific theory that human activity, through the increasing emissions of greenhouse gases, is accelerating the warming of the Earth's atmosphere. After more than a century of research and decades of debate, the science converged towards the view that this was a credible thesis. Global temperatures have continued to increase at a rate that exceeds historical records of natural climate variation. The reinsurance industry (which essentially provides insurance for insurance companies) realised that this credible thesis was increasing the frequency and scale of risks they were managing, which meant they were more expensive to insure. The global capital market worked out that if climate risk increased costs, then investments exposed to this risk became more expensive, and therefore more marginal. This process of steering away from high-greenhouse-emission activities has been forged in the engines of capitalism, guided by a series of dispassionate arguments made over decades by experts in their respective fields. And yet some of us, including a persistent clique of conservative politicians and commentators, continue to dismiss their advice.

In Australia, the growing awareness of climate change meant maximum disruption. Globally, the three biggest sources of human-produced greenhouse gas emissions were electricity generation, transport and agriculture. Australia was – and is – a per-capita world-leading emitter in all three. Making deep and rapid cuts in greenhouse gas emissions would require a much greater response, proportionally, from countries like Australia than from others.

The effect of climate change on Australia's electricity system was like kryptonite on Superman. Its strength was now its weakness. The big, brutalist coal-fired generators sitting atop hundreds of years of cheap fuel were now at risk. Australia found itself long on high-emissions electricity, and pretty short on everything else. The disruption by climate change coincided with the evolution of an increasingly volatile electricity system in Australia. Energy economist Paul Simshauser identified that the effect of rising wealth in Australia was driving increased peaks in electricity demand to heat and cool large houses, while the cost of supplying these extra peaks in demand was also rising sharply. It was going to be more expensive, even without climate change.

Other developed countries also had to pivot significantly to reduce their emissions, but the relative scale of their challenge was more modest. Their electricity renovations would require some replacements, but most had a bigger share of medium- (gas) and zero-emissions (nuclear or hydro) generators – or they were connected to a neighbouring country that could help out. One of the big responses to climate change in Europe has been to look at increasing international transmission lines into France, which runs almost entirely on nuclear power.

Australia had no one to connect to. It also had relatively little zero-emissions generation; hydro was only around seven per cent of total supply and could not be increased. The age of Australia's coal-fired generators meant they were never going to justify being retrofitted with technologies to capture and bury the greenhouse gases, even if that technology ever proved to be viable. In short, Australia was the only country in the world faced with the real prospect of having to almost completely rebuild its electricity supply from scratch.

Electricity gets political

Politics is the antithesis of science. Science is rational, incrementalist, impassive. Politics is irrational, absolutist, emotive. Politics, by necessity, prioritises populism over evidence, the short term over the long term, symbolism over substance. At the start of the 21st century, the developed world's political classes had to consider how to manage the impact of a large and complex scientific risk on a large and complex industrial machine. What could possibly go wrong?

The global response was chaotic. International voluntary agreements (like the Kyoto Protocol) with mild, non-binding targets were drafted. Some governments, such as Australia and the US, initially refused to sign but then met the targets anyway. Mostly emissions were pushed up and down by external events: China built hundreds of coal-fired power stations to fuel its rapidly expanding economy, then started to close those nearest its big cities because of smog. The US delivered the biggest emissions reductions of any developed country, not because of climate policy, but because it discovered cheap unconventional (shale and coal seam) gas. Japan's emissions increased sharply after the Fukushima nuclear accident in 2011 as they ran their coal and gas plants harder to cover the closure of dozens of nuclear power stations.

Europe moved early and confidently to launch an emissions

trading scheme in 2005, only to basically turn it off after the global financial crisis in 2008. Some countries, including Denmark, increased their renewable energy generation. Germany also built new wind and solar power stations, then decided to close all its nuclear power stations after the Fukushima accident. It ended up building new coal-fired power to fill the gap. For all the billions of euros spent, its emissions remained largely the same.

In Australia, the conservative Howard government decided to ignore climate change, successfully at first. At the start of the 21st century, the issue posed little real political threat. A handful of modest renewable schemes were implemented. Elections in 2001 and 2004 largely ignored the issue. Prophetically, what transformed the conversation around climate change in Australia was the changing climate.

In 2006, southern Australia was gripped by the worst drought since European settlement. Dam levels began to plummet, and by the year's end, brutal water restrictions were imposed on the residents of major cities and regional towns. Lawns could not be watered and were left to die. Pools could not be filled. The suburban backyards of millions of Australians were becoming dustbowls. Around this time, former US Vice President Al Gore released *An Inconvenient Truth*, a film about the threat of climate change. It was later proven to contain some significant exaggerations and factual errors, and not many Australians went to see it, but it contributed to the escalating Zeitgeist. Suddenly climate change wasn't just a scientific theory: it was a thing, happening in Australia, right now.

Global business was shifting too. The reinsurance industry had been managing increasing insurance payouts on more-frequent and more-intense weather events. The rising business risks of a warming planet drove a complete rethink on the issue. Businesses didn't engage with scientific debates, but they knew all about risk management. By the end of 2006, this reset had filtered through to global lending and financing. Companies added climate change to their risk-management strategies, and the Business Council of Australia was calling on the federal government to take decisive action on climate change. After a decade of carefully ignoring the issue, John Howard was blindsided by his traditional allies.

Politicians began scrambling to outbid each other: Howard called for emissions trading; his new opponent, Kevin Rudd, called for a suite of climate policies including emissions trading, renewables targets, ratification of the Kyoto Protocol and subsidised home insulation schemes. Neither leader understood the detail of how the electricity system worked or how it would respond to their pop-up policy measures. That wasn't the point. Labor won the 2007 election in a climate change–fuelled landslide. The result politicised climate and energy policy in Australia. It also laid the foundations for the ideological divide on energy policy between conservative and progressive Australia.

After the election there was climate policy détente in Australia, which was then shattered in 2008 by the fallout from the global financial crisis. The centrepiece of global commitment was supposed to take place in December 2009 at the United Nations Climate Change Conference in Copenhagen. It dissolved into a wake. Conservatives who were against climate change action reasserted themselves. They turned on Australia's efforts to mitigate emissions of greenhouse gases as either unnecessary, trivial in a global context or an expensive folly: bringing higher energy costs and job losses, a reckless rush ahead of the rest of the world to embrace dubious science. Coal became the poster child for conservatives; big renewables targets became the mascot for progressives.

Having politicised climate change, state and federal governments did what they do best: they spent money and announced things to impress and appease their constituents. They promised large, uncosted and untested renewables targets, onerous then generous planning conditions on wind farms, over-generous subsidies for solar panels, commitments to build new government-funded coal-fired power stations, carbon trading, carbon taxes, repeal of carbon taxes, gas moratoriums, giant batteries and giant transmission lines. Meanwhile, most voters just wanted to know three things: had they sorted this climate thing? Would power bills come down? Would the lights stay on?

This revealed the bleeding obvious: most politicians, in fact most Australians, didn't know how the electricity system worked. For a century they hadn't needed to. Governments knew what they needed to know: cheap electricity made voters happy, blackouts made them unhappy. The answer was to get smart electricity people to roll it out and make it work. When it didn't, move bloody hell until it did again. Climate and energy policy in Australia needed to be visionary, strategic, bipartisan. Instead we got handed a set of university-politics steak knives to solve the biggest structural reform challenge of a generation. Only this time we were not running the student union: we were shaping the very future of the economy. We entered a slow but lethal death spiral. We're still in it.

How to break an electricity grid

Electricity is a rather unique commodity: it can't be stored at grid scale. The amount of electricity being produced to supply a grid is constantly changing, every second of every day. As lights, air conditioners and factory machines are switched on and off, generators make constant adjustments for the amount of power demanded. Managing this constant variability is the main game when it comes to running electricity grids. Demand can vary wildly, skyrocketing during heat waves and cold snaps, and falling down to very low levels in the middle of the night and during holidays. It means electricity systems need to be durable and adaptable, and have a range of options to find power at the oddest of times and circumstances. When you are running 24/7, the unexpected happens.

The two things an electricity grid must be able to do are, one: have enough capacity to meet those periods of peak demand, and, two: constantly manage the quality of the power in the grid so it is operating at a safe level. For the past 100 years, at the core of every grid there have been large power stations that could track the amount of power in the system and automatically regulate their output to keep it stable. The key measurements for this are written on the back of every appliance in your house: voltage and frequency.

In Australia we run a 230-volt system with 50-hertz frequency. These are the key measures of what is called 'power quality'. The electricity in the grid is like a car driving down a road. The amount of pressure on the accelerator represents the voltage (how much power is in the grid), and the steering wheel is the frequency. It has to stay pretty straight on the road at 50 hertz or the car will crash. The voltage can vary slightly around 230 volts, so long as the car doesn't go too fast or too slow.

This is easy to manage when you have large, controllable generators that can dial up and down to keep the grid inside the safe limits. It gets harder when you start to replace them at scale with technologies such as wind and solar, which are less controllable and don't run all the time. This switch over is possible, but it requires careful planning to ensure all the tasks of the controllable generator are being adequately replaced.

The decade-long political tug of war over climate – with the electricity system as the rope – ignored all of this. The system started to be pulled in different directions. Building 21 new wind farms in a small grid like South Australia's had the effect of forcing the state's last coal-fired power station out of business a decade ahead of time. Environmentalists cheered, but suddenly the state's ability to both meet capacity and manage power quality was significantly diminished.

Installing 2 million rooftop solar systems on households across Australia had an effect too. The subsidies came from higher power bills for non-solar households. In some places it loaded up the network of poles and wires with more power than they could move.

Committing to an emissions trading scheme (ETS), cancelling it, and introducing a carbon tax, then repealing it, also had an effect. New gas power stations that were built in expectation of the ETS no longer looked like good investments by the time they were finally commissioned. As a result, future investors needed to 'see the money' before they'd bankroll the new generators needed to replace the old high-emissions power. Without that stability, the bankers' chequebooks stayed closed. Plans for new capacity were shelved. Only renewables kept being built behind the bipartisan Renewable Energy Target (RET). Renewables are a key part of the new replacement grid, but building a grid using only renewables is like building a brick wall without mortar.

Since 2012, ten old coal-fired power stations have closed in Australia. The remaining stations that were part of the postwar boom are also approaching retirement. Over the next 15 years around half of them will close. And we still don't have a plan to replace them. Political leaders have, instead, just doubled down on the politics. Blaming each other, blaming the industry, privatisation, deregulation or anything else they can think of. And every day this continues the crisis only deepens, and the risk of serious and permanent harm to the economy intensifies.

What next?

The clock is ticking. The current fleet of coal-fired power stations was built from the early 1960s. They reach the end of their operating lives at around 50 years. So, to put it bluntly, we are gradually running out of electricity. In the past decade we have built new wind and solar generation. This will help. But wind and solar on their own are not a like-for-like replacement for coal. We need to add mortar to the bricks and use other technologies to supply the grid after the sun has gone down and the wind has stopped blowing.

So what will our new electricity grid run on? What will it look like? Some argue we should just build cleaner coal-fired power stations to replace the old ones as they exit. New high efficiency, low emissions (HELE) coal-fired generators would just be an upgrade of existing coal-fired generators. They would still produce relatively high greenhouse gas emissions and they're expensive. And regardless, there are also no signs that banks and businesses are interested in building these, considering the unknown carbon risk. A government might, but it would take seven years to get built, and governments can change two or three times in that period.

Australia can and probably will use more gas. Gas produces half the emissions of coal and is more flexible, making it much more suited to working with renewables. It's also more expensive and unhelpfully scarce in eastern Australia. Large batteries to back up renewables are evolving and are already proving to be effective in managing power quality, but they are still relatively expensive and do not store the volumes needed to power the grid for long periods of time. Pumped hydro may help with storing bulk power – this is when water is pumped from a low to a high elevation, then used to create hydroelectricity – but cost and scale will be challenging. There are useful and important changes we can make in the way we use electricity to better match the capacities of the new generation sources. One day we might even need to talk seriously about nuclear energy.

The way forward is not clear, but nor is it impossible. Current estimates suggest we will need more than 100 billion dollars of new investment in new generation by 2050, and about the same on the network of poles and wires. That is a lot of money. If we want businesses rather than governments to do the building, we will need to give them the conditions to bankroll their new assets. If not, then governments will need to step in and divert funds from elsewhere – maybe schools and hospitals, roads and services.

Electricity in Australia – and globally – has begun a transformation that will continue for decades. Electricity is likely, but not certain, to power a lot more cars and other transport vehicles over the coming decades. Our houses and buildings are likely to generate and store more electricity and will also use less, more flexibly, to do the same job. This will involve smart, automatic systems and devices that reflect the computer age we live in. Appliances in millions of homes will switch on and off, up and down like a giant electricity orchestra. The system will breathe in and out with changes in wind and sunshine, but it will still be anchored around a network of power stations. The big question is what will power these.

This is a sophisticated, high-tech transformation that will require mature, strategic leadership, and coordination that rises above political theatre. Ultimately the thing that makes electricity so useful is that we don't have to think about it. The measure of a successful transformation will be whether electricity returns to its rightful place in the background of our lives. Success means we stop talking about it. It just works.

What is a watt?

A watt is a measure of the rate at which energy is consumed. Every appliance in your home is rated in watts (W). A kilowatt (kW) is 1000 watts, and a megawatt (MW) is one million watts. The number of watts describes how much electricity that device needs in order to operate. A 60-watt old-fashioned incandescent light bulb needs 60 watts. A hair dryer might need 800 watts, a microwave 1000 watts. Generation is measured in watts too. A typical rooftop solar photovoltaic (PV) system might produce around 4 kilowatts (4000 watts). That means at its maximum output it will produce 4 kilowatts when the sun is brightest. The generation capacity of a large coal-fired power station is also measured in watts. The biggest generator in Australia is the Eraring Power Station near Newcastle in New South Wales. Its maximum capacity is 2880 megawatts (MW), or 2,880,000,000 watts.

Wind and solar farms are also rated in watts. A typical new wind turbine in Australia will have a maximum output of around 2 to 3 megawatts. Wind farms that consist of multiple wind turbines can produce anywhere from 25 megawatts up to the 420 megawatts that the Macarthur Wind Farm in Victoria generates. Solar farms can produce a similar amount.

The quantity of electricity consumed is called energy. The longer an appliance is connected the more energy it consumes, thus the unit of energy is the power in watts multiplied by the duration of operation. The units of energy in the electricity industry are watt hours. So a 60-watt light bulb left on for an hour would consume 60 watt hours. Electricity is charged on retail bills by this measure: the kilowatt hour (kWh). The average Australian household consumes around 16 kilowatt hours of electricity every day. This tends to increase dramatically in heat waves and cold snaps. At times of high demand such as hot or cold evenings, a household might consume that amount of energy in the space of a few hours. The demand often falls to much lower levels overnight.

Finally, to give a sense of the scale of electricity generated by a power station, it is common to measure it in the invented unit of 'numbers of houses powered'. Because the wind does not blow at maximum all the time, during the course of a year a 100-megawatt wind farm will produce about 30 per cent of its rated output – this is called its capacity factor. So a 100-megawatt wind farm (100 megawatts) × 8760 (number of hours in a year) × 0.3 (capacity factor) = 262,800 megawatt hours produced per year. The amount of electricity consumed on average per household each year is 16 kilowatt hours × 365 days = 5840 kilowatt hours, or 5.84 megawatt hours. So the average number of houses powered by the 100-megawatt wind farm is 262,800/5.84 = 45,000 houses. This is a very imprecise measure, as in reality a number of different power stations will power that household. And neither household demand nor power station generation is constant. But at least it gives you an idea.

1

WHERE DOES ELECTRICITY COME FROM?

Modern life is complex. Within living memory we have come to rely on an expanding cohort of technologies to help us move around, see at night and communicate, to inform, to entertain and to simply get things done, but most of us have little idea how these technologies actually work. Nor do we need to know. They are just services or devices constantly evolving in the background. If something goes wrong, we either get an expert to fix it or we replace it.

The grandparent of all of these modern technologies is electricity. Understanding how electricity actually works wasn't a priority for the average person. If the power went out, we just waited in the dark until the problem was sorted out.

This relationship with electricity changed at the start of the 21st century, driven by two factors. Consumers started becoming an active part of the machine, initially by making and selling electricity back into the grid with solar panels, then by dialling demand up and down, and more recently by storing electricity in batteries and discharging them at different times. The second big game changer was climate change. It required a radical change to the future design of the electricity system. Change on this scale invoked a heated debate. Electricity became a political issue in its own right.

For the past decade, Australians have been asked to form opinions about radically different approaches to the design of the electricity grid in the 21st century. And most of us haven't a clue what anyone is talking about.

So if we are going to have a public debate about the design of electricity in the 21st century, it's probably a good idea to understand where this machine came from, and how it came to play such a central and ubiquitous role in our lives. Understanding how we got here may help guide how we redesign electricity system in the 21st century and beyond.

Discovering electricity

The process of discovering electricity was a 300-year-long grope in the dark. For centuries, electricity remained a mysterious force, and its discovery was largely due to a process of trial and error. Experimentation and explorations were almost entirely observational: early pioneers slowly unravelled the mysteries of electricity by observing how it acted, rather than by developing a body of theoretical science and then testing the results in the field (as with, say, nuclear fission). The science to explain how electricity worked came years later. The electron, which sits at the very core of electrical science, was not discovered until 14 years after the first grids were being rolled out in London, New York and Melbourne in 1882.

Electricity appears in the natural world in two ways: lightning and static electricity. The word 'electricity' is derived from the Greek word for amber, *electron*. The ancient Greeks first observed that pieces of amber had an attractive force when rubbed. The term 'electricity' was coined in the 17th century after the Greek observations were replicated to describe materials that attracted others. Static electricity was easy to make and shocking to touch, and it quickly intrigued both scientists and the general public. Static electricity generators – aka giant friction machines – were used mainly as sideshow entertainment, where sparks and lights were created while patrons, and sometimes other less animate objects, were shocked into motion with an invisible electrical jolt.

Attempts to explain the science behind electricity floundered. There was simply too little known about physics and chemistry to make sense of the invisible movements being created. It wasn't until the early 18th century that, by accident, it was discovered that a static electricity charge could be moved via a conductor. This led to the famous experiment by Philadelphia printer Benjamin Franklin in 1752, in which he flew a kite in a thunderstorm and conducted the charge down the string to electrify a key. This led scientists to think about electricity as a way of moving this charge.

Electricity as a machine for service was ultimately a product of the steam age and the scientific revolution, unravelled by some remarkable minds: Galvani, Volta, Ampère, Faraday, Ohm, Edison and Tesla among them. Many of these familiar-sounding names have been immortalised in the modern-day electricity vernacular as reward for their contributions. Electricity's transformation from curiosity to industrial powerhouse was made possible by four key discoveries in the 19th century: the battery (a constant flow), light (the first mass marketable application for its use), the generator (factory production of electricity) and the transformer (helping to move large quantities of electricity around).

Batteries

The old static electricity machines took various forms, but generally they comprised a glass or metal disc or ball that stored the static charge, and a hand-cranked mechanical device to create the friction that produced the charge. These devices were often bulky and not particularly easy to move around. Naturally, there was some interest in finding a way to store this charge. Given no one actually knew what static electrical charge was, some theorised it might be a liquid, which led to a range of experiments attempting to store the charge in jars filled with water.

The first battery made to store this static charge was called a Leyden jar (named after Leiden University, where it was created). The first Leyden jar was a glass jar lined with metal foil and half-filled with water, with a chain dangling down to contact the metal foil. This chain was then hooked up to a source of static electricity. But despite repeated attempts, the jar didn't seem to store an electrical charge.

Reflecting the trial-and-error nature of electricity discovery, one day in 1744, the Dutch physicist conducting the experiment, Pieter van Musschenbroek, accidentally held the jar while it was charging. He then touched the top of the jar and received a nasty shock. The water was irrelevant. All that was required were two conductors (his hand and the metal foil) separated by an insulator (glass), which stored opposite charges until they were connected. Leyden jars were then used to administer controlled electric shocks as a 'medical treatment' for anything from constipation to paralysis, as well as providing a more convenient way of performing the usual array of sideshow-alley tricks.

While completely fraudulent from a medical perspective, this development did steer electricity research towards the emerging schools of biology and physiology, which, perhaps unexpectedly, led to further development. In 1780, an Italian physiologist from Bologna called Luigi Galvani was exploring the idea that electricity was some sort of life force. He found he could induce spasms in the legs of dead frogs when he zapped them with static electricity. He later found (completely by accident) that he could also induce spasms simply by touching the frogs with different metals.

Such was the fascination and mystery surrounding electricity at the time, the idea it was some kind of life force quickly took and held the public's imagination. This is why, in 1816, Mary Shelley used the idea as the central scientific construct in her infamous gothic horror novel, *Frankenstein*.

After Galvani's death, his student – and rival – Alessandro Volta continued to investigate the phenomenon. Volta was unimpressed by Galvani's theory. When repeating his rival's experiments, he discovered that the movement in the dead frogs' legs was caused by the different properties of the metals used, not some life force. He then got rid of the dead frogs altogether and went about building a battery. Volta found he could create a constant charge by stacking alternating layers of zinc and silver, separated by paper soaked in salt water. The zinc dissolved into the salt water, releasing electrons and creating a constant electrical charge.

This was the world's first chemical battery, the voltaic pile. The invention of the battery preceded the electricity grid by nearly a century. If there was a defining moment in the exploration of electricity, this was probably it. Volta's battery literally changed everything. It enabled further scientific breakthroughs in chemistry and electrical engineering, and within decades was used to power new world-changing technologies: the telegraph and the telephone.

Generators

Chemical batteries continued to evolve after Volta's breakthrough, but they only produced a trickle of energy. This was useful for laboratory experiments or in technology where only small amounts of current were required, such as telephones. The development of industrial- and utility-scale electricity would require a significant scale-up in the quantity of power created.

The invention of the generator was made possible by another accidental observation: this time of the link between electricity and magnetism. Parisian scientist André-Marie Ampère realised that electrically charged wires created a magnetic field. Then an Englishman, Michael Faraday, discovered that the reverse was also true: magnetic fields could create electric currents. In 1831 he built a rotating coil of conductive wire within the lines of force of a magnetic field. When the coil spun, the magnets released electrons in the wire, which created an electromagnetic force. The Faraday disc was the world's first electric generator, known as a magneto.

Faraday's discovery produced an alternating current (AC), meaning the current switched direction with each half-rotation of the magnetic wheel inside the generator. Electricity produced by chemical batteries is a direct current (DC), flowing continuously in one direction (because it has nothing rotating), and since batteries were the main game at the time, the magneto was initially considered useless.

The discovery and development of electric generators quickly evolved into the invention of the electric motor, which is basically a generator operating in reverse. Electric motors were promising, but their usefulness depended on how much power they could access. Bigger motors needed bigger generators. The operational mirror between motors and generators revealed that electricity was, in its simplest form, a system for converting and moving energy.

Light

The first big commercial use for electricity was the telegraph. Batteries were soon able to provide enough current to transmit electrical flows by wire over long distances. Logically, this flow could also be interrupted. Coding the interruptions enabled messages to be sent. Various telegraphic systems were developed during the early 19th century, but by 1837, a simple but effective system of coded dots and dashes designed by Samuel Morse and Alfred Vail became the standard. In Australia, the telegraph revolutionised communications, enabling information from mother Britain to be received in minutes instead of weeks. By 1872, Australia could receive news as it happened via telegraph from the rest of the world.

Electricity delivered information, then light. Until the middle of the 19th century, the world's cities were mostly filthy, smelly and dark. They were dangerous and unpleasant places, covered in soot from coal, and stinking from open sewers and worse in the streets. And they were dark. The ability to light the darkness in streets, homes and workplaces was a commodity of immense value. Reticulated gas lights were the first form of public and private lighting. Sydney's streets were lit by gas from 1841 by The Australian

Gas Light Company (now known as AGL). Gas lighting was also installed in some wealthy homes and businesses through the middle of the 19th century.

The first types of electric lights were called arc lamps. They produced a furiously bright industrial-scale light by running a large (1000-watt) current between two carbon-charged electrodes kept slightly apart. These were basically giant torches: a huge lantern attached to a huge battery. Some uses were found for these arc lamps – lighting mines, factories, streets and even some theatres – but they were limited: they blew quickly, were expensive to make and run, were far too bright for most applications and drew huge amounts of power.

Aside from their more practical uses, arc lamps became incredibly popular as public displays and for celebrations. Most reports of 'first electric light' were public arc lamp 'shows'. Their popularity as entertainment led to furious competition to make a gentler, more useful light. In 1879, three similar patents (including that of US inventor Thomas Edison) were lodged for the same technology: an incandescent light bulb that drew less power and glowed less brightly. Light bulb 1.0 used the same basic design that we use today, although the first bulbs only had a life of around 13 hours.

The grid

Armed with a working light and generators to supply regular electricity, Edison and his competitors rushed to roll out electricity networks to businesses and homes. It was a fierce contest. Edison built his first grid in London and a second in New York, servicing just 53 dwellings in the Wall Street financial district. The use of DC power limited the size of the grid significantly.

DC electricity is commonplace today. It is used to run small appliances and systems: car electrics, laptop computers, batteries, electric torches, campervans, and even some small, discrete or remote grids. It is also the way electricity is generated by solar cells, and the way it is stored in batteries. The problem with using DC in a large grid in the 19th century was that it operated like a single, low fixed-gear bicycle; it couldn't vary (scale up) the voltage for different uses. This was limiting: moving electricity over longer distances required increasing the voltage (or 'the pressure' as it was referred to within the industry) to avoid load losses. It also required a fairly distributed grid, with power stations located among consumers, rather than at the outskirts.

As grid ambitions got bigger, larger volumes of electricity needed to be moved greater distances more efficiently. This meant voltage (pressure) needed to be scaled up and then scaled down to different levels for different uses. Flexibility was key. While households could all be run on the same voltage, factories and other commercial applications were realising they would need a range of different voltages for their motors, refrigerators, pumps and smelters. A modern electricity grid needed lots of different gears.

To solve this, engineers went back to Faraday's magneto, previously considered useless because of its alternating current. AC enabled electricity to be moved around a large grid, via the invention of transformers. At its simplest, a transformer consisted of two coils placed next to each other, but without touching. Using what is called electromagnetic induction (not unlike an induction cooktop), the charge in one coil transferred its energy to the other, magnetically inducing a different voltage. Each of the two coils had a different number of loops, and the difference between these determined whether the voltage increased or decreased. Importantly, this gear change was only possible using AC.

Today, transformers are everywhere. Small ones are in the box embedded in power cords for computers and mobile phones, large ones are located through the electricity grid at substations and on transmission lines. They act as a series of locks in a canal, bringing voltage up and down, enabling high-voltage transmission of electricity over long distances with very low losses and lower-voltage distribution of electricity into our homes.

At the end of the 19th century, the electricity grid had a clear, uniform design and shape: it used AC; it was powered by steam (and some hydro) generators and used transformers to shift voltages in order to move energy more efficiently over large distances; its network of dangerous power lines needed to be run on poles to keep it safely away from human contact; and it powered a growing range of lights, motors and other devices into a growing number of businesses and households. And anyone who knew about it wanted to be connected.

Electricity in Australia

Electricity was the IT boom of the late 19th century. Every government wanted it, every entrepreneur wanted to sell it. News of the take-up of electricity in the US and Europe reached the young colonies in Australia almost instantly via the electric telegraph. It was as thrilling here as anywhere else. Australia's remoteness did little to dampen local enthusiasm – the country was electrified in the time it took to get the required machinery on ships and installed once it landed.

The hype started with a scattering of brilliant arc lamp displays in public places: in 1863, Observatory Hill in Sydney and Parliament

House in Melbourne were lit to celebrate a royal wedding. But the real coming of age for Australian electricity was in 1879, when half-a-dozen arc lamps were used to light the Melbourne Cricket Ground for two night games of Australian Rules football. (It says a lot about Australians that we used electricity to organise a football match under lights before we'd even built a grid.)

The gold rush framed the culture of Melbourne in the late 19th century. The economic boom from gold brought with it lots of new money and a more 'adventurous' business culture. Like Wall Street in the 1980s, Melbourne of the 1880s was awash with cash and keen to spend it. As a result, Melbourne became one of the first cities in the world to install an electricity grid.

In 1881, one of the first electricity start-ups, the Victorian Electric Light Company, won a Melbourne City Council electricity contract to light the (now defunct) Eastern Market. They soon discovered that building electricity infrastructure from scratch was a capital-intensive business. Within 12 months, the Victorian Electric Light Company had to be re-floated as the Australian Electric Company, and they used the new funds to build a small brick building in Russell Place (just west of the then market on the corner of Bourke and Exhibition Streets) in the Melbourne CBD. The ground floor of this inauspicious building housed the first power station in Australia.

Commissioned in 1882, this small grid powered up at the same time as Edison's landmark grid in Manhattan. Two small steam engines were used to turn a DC dynamo that powered the nearby market. They also sold street lighting in the immediate vicinity and picked up some local commercial and residential customers, including incandescent lighting for some of the nearby theatres. But it was a strictly local affair. As in Manhattan, Melbourne's new DC grid did not have the voltage to reach more than a few hundred metres from the generator. In five years, the feeble DC plant was replaced with a larger AC generator. The site today is still used as a substation by the local network business CitiPower.

Gas businesses had been supplying their own duller form of street lighting in Australian cities since the 1840s and were less-than-impressed about their incumbency being challenged by this new technology. AGL was big in Sydney and big in gas lighting. It wasn't going down without a fight. Faced with the threat of electricity, 19th-century AGL cut its street lighting fees, recontracted customers, upgraded its lamps and leaned on the political process in an attempt to stave off electric lighting. As a result of these efforts, Sydney didn't get electric street lighting until 1904, long after places such as Tamworth (1888), Port Adelaide (1889), Penrith (1890), Broken Hill (1891) and Launceston (1895). Nearly a century later, AGL finally diversified into electricity.

Electric street lighting was immensely popular, quick to install, publicly accessible and relatively affordable. Soon another electricity business emerged: trams. At the turn of the century, Melbourne and Sydney had a population of around half a million people each. The distances across town were becoming further than walking could convenience. Councils began to explore the idea of providing public transportation: carts and trolleys pulled by horses, wagons pulled by cables using a steam engine, or even carriages powered by their very own steam engine. Trolleys fitted with electric motors soon emerged as a serious option.

The first electric tram service in Australia was built in Hobart. It was shipped in from Germany in 1893 like a giant train set: double-decker tram carriages, tracks and electricity supply (a coal-fired powerhouse that was the first electricity supply in Tasmania). The modest but extremely popular tram service ran north and south of central Hobart along the coast. Like many early electricity start-ups, the creatively named Hobart Electric Tramway Company struggled commercially. Even though the trams were well patronised, the city's relatively small population meant they simply couldn't earn enough money to cover the up-front cost of building the giant train set, and it was eventually bought out by Hobart Council in 1912.

Through the 1880s Sydney had developed a small network of horse-drawn, cable and steam-powered trams. Electric trams were trialled in the eastern suburbs between Bondi and Coogee and employed full-time from 1896. To power them, a new generator was built at Ultimo. The building still stands today, better known as the Powerhouse Museum. Electric trams were soon introduced in every major Australian city: Brisbane in 1897, Perth in 1899, then Melbourne's first commuting (and privately owned) electric tram to Essendon began in 1906. Adelaide completed the set in 1909.

To power these growing networks of street lights and trams, electric generators were built near or in the middle of town. Two companies built generators in the industrial Melbourne suburb of Richmond in 1891. Melbourne City Council built its first generator on Spencer Street in the CBD in 1892, and the site was still generating until 1982 (it's now an apartment building). Adelaide's first power station was built in the CBD on Grenfell Street (now converted to an Aboriginal cultural institute). Brisbane's first power station was sited behind the General Post Office (now office buildings). Sydney had power stations ringing the city: Pyrmont (now The Star casino), Ultimo (Powerhouse Museum), Balmain (apartments), Rozelle (derelict White Bay Power Station still standing) and Port Botany (container terminals). Even the first Tasmanian hydro power station – Duck Reach, built in 1895 – was sited on the Esk River just on the outskirts of Launceston (it's now a museum).

State by state

The early pioneering of electricity in Australia was driven by a bunch of private sector start-ups and enthusiastic local governments. Private electricity entrepreneurs were keen to cash in on the promise of fast growth and big returns, only to find the capital demands of building and running electricity utilities and transport businesses were more than they had bargained for. The first electricity companies were continually relisting, refinancing, merging or just going out of business from trying to keep up with the escalating commercial demands of building large generators, connecting customers and then recovering these costs by billing their growing customer base.

This erratic start soon frustrated their impatient government contractors, who wanted to deliver the marvel of electrification faster and wider. By the start of the First World War, electricity powered most major-city trams, and electric street lighting had spread to some suburbs and larger towns. The most lucrative customers were businesses: factories that used electric motors and lots of electric lighting. Residential customers were being connected at a slower pace, still more of an afterthought. Connection was subject to the proximity of existing supply and the capacity of the electricity company in question to manage small accounts. Most early-20th-century electricity companies were generators first, retailers second.

There were technical problems emerging too. Electricity grids were evolving into a patchwork of local generators and a web of power lines radiating out from them with different operating systems. Many suburbs and towns still operated their electricity grid in complete isolation. They often ran on different voltages and other operating requirements. This meant motors and appliances couldn't always be used nationally. By the early 20th century local councils took matters into their own hands and began aggressively taking over the private electric companies.

These rationalisations created larger grids. The cheapest and safest way of moving electricity around a city was to hang power lines high on street poles. A network of substations was built: secure buildings used to step the power down for household and commercial use. Many of these substations from the early 20th century are still in operation today. The oldest of them can be found scattered through the now-affluent inner suburbs of Australian cities. They were often built as a celebration of modernity with an eye for architectural merit, some in the style of Californian bungalows or Spanish missions – brutalist red-brick, like churches or town halls. Many of us will have walked past or vaguely admired these odd, slightly grand windowless buildings at some time in our recent past.

In Victoria there was an additional problem emerging: not enough coal. By the end of the First World War, Hunter Valley coal out of Newcastle was supplying the early power stations in New South Wales, Victoria, South Australia and Western Australia. Chronic industrial action in the New South Wales coalfields created shortages, which in turn affected electricity supply. By 1920, there were calls in Victoria to move toward self-sufficiency: electricity was considered too vital to the state's development – and besides, the Victorians hated relying on New South Wales for anything.

The Victorian Government eventually stepped in, deciding to exploit the vast brown-coal (lignite) fields in the Latrobe Valley. Brown coal was wetter and had less energy than the world-class coal from the Hunter Valley, but at least it was local. It was also more expensive to move brown coal long distances, so the new power stations would need to be built at the mine mouth, and the electricity transmitted 160 kilometres to Melbourne. It was a big undertaking, requiring a new single government utility to deliver it: the State Electricity Commission of Victoria. Its first chair was iconic Australian Sir John Monash, and by 1924 it had built its first brown-coal-fired power station at Yallourn. By 1934 it had completed the acquisition of the last remaining private electricity company in Victoria.

Other states were either already there or soon followed. In Tasmania, a private attempt to power a zinc smelter using a hydro generator in the Central Highlands' Great Lake went broke in 1914. The popular but struggling Hobart tramway had already been bought out by Hobart Council in 1912. Both the trams and the Great Lakes project were acquired by the new Hydro Electric Commission in 1916. Hydro quickly proved to be cheaper and more efficient than thermal generation, and Hobart's coal generation was decommissioned soon after.

State Electricity Commissions were created in Queensland in 1938 and Western Australia in 1945. The Electricity Commission of New South Wales was formed in 1950. In 1946, South Australia followed a similar path to Victoria's when then premier Tom Playford borrowed money from the Chifley government to buy out the private monopoly Adelaide Electric Supply Company after it refused to source its coal from the state's brown-coal fields at Leigh Creek. Playford transferred the assets to the newly created Electricity Trust of South Australia, and then started building new brown-coal-fired power stations at Port Augusta.

Postwar boom

By the start of the Second World War, most urban Australian dwellings had electricity. Demand was modest: electricity was used mainly for lighting, and possibly powering a radio or record player. In 1946 only 16 per cent of Australian households had a refrigerator. Only two per cent had a washing machine. Air conditioning was unheard of; water was heated by gas or solid fuel. Australia's unrenovated pre-war houses were notorious for their lack of power points. Domestic life was labour intensive. The increased participation of women in the workforce was held back by the scale of work they were expected to do at home.

The war had severely constrained the development of Australia's electricity systems. But this changed fast. As the global economy returned to normal, post-war industrial production shifted to a new range of consumer electrical goods: washing machines, refrigerators, toasters, irons and then televisions, air conditioners and hot-water systems. The post-war boom drove new investment in resources and manufacturing, which required more electricity. Rapid escalation of electricity supply had become critical both to the political survival of governments and to the continued growth of the economy. The scale of new generation required meant that most of the centrally located power stations were too small and too close. The solution had been signalled in Victoria: build new, much larger power stations out of town.

Australia's current fleet of coal-fired generators are the result of this dramatic post-war infrastructure boom. South Australia built at Port Augusta, near its brown-coal resources at Leigh Creek; Victoria expanded capacity in the Latrobe Valley; New South Wales located power stations through the Hunter Valley; Queensland, through the Darling Downs and west of

Rockhampton; and Western Australia, around its Collie coalfields. One of the biggest challenges for the new state-owned electricity monopolies was keeping up: electricity demand–supply balance was already tight at the start of the boom, and demand for more electricity grew rapidly for the next four decades: an eightfold increase between 1955 and 1995. The lifespan of many of these older power stations today is informed by how hard they were run (or in some cases, flogged) in their early years.

Ministers and premiers rarely understood the detailed workings of these giant electricity machines; they just demanded fast growth, low prices and increasing reliability. From that perspective, a coal-fired electricity system made complete sense. Cheap electricity was enabled by two things: the discount rates at which state governments could borrow to build power stations, and access to cheap fuels to power them. And coal was as cheap as it got. Brown coal looked like muddy wood and cost about as much. The black coal used in Australian power stations was of the lowest (and cheapest) grade mined in New South Wales and Queensland. (The higher quality coals were reserved for export.)

In the second half of the 20th century, the most controversial electricity generation projects weren't coal-fired but renewable. The biggest electricity generator in Tasmania was, and still is, the 432-megawatt Gordon Power Station. It came on line in 1978 after the Hydro Electric Commission (HEC) flooded Lake Pedder in the Southwest National Park, despite fierce opposition from conservation groups. In 1978, the HEC proposed the building of the Gordon-below-Franklin Dam to power an extra 180 megawatts of new generation. The proposal divided Tasmania and became a hot-button national issue. Cars all over the inner suburbs of Sydney, Melbourne and Adelaide sported the famous

'No Dams' campaign sticker. The dam became a major issue in the 1983 federal election, which was won by Labor. The new prime minister, Bob Hawke, duly took the matter to the High Court, which ruled that the federal government had the constitutional right to overrule the Tasmanian Government based on Australia's international obligations under the World Heritage Convention. This protest movement launched the Greens as a political movement in Australia – in other words, the Greens were formed in opposition to renewable energy.

By contrast there was little uproar about the other big hydro project in Australia: the Snowy Mountains hydroelectric scheme. Built between 1949 and 1974, it remains the largest engineering project ever undertaken in Australia and has an iconic status in the Australian identity. The Snowy hydro scheme used a network of dams, lakes, pumps, tunnels and pipelines to divert more than two thousand billion litres of water each year west into the Murray–Darling Basin. This diversion increased the amount of water available for irrigation while enabling more than 2000 megawatts of hydroelectric generation. In order to build it, three towns within the Kosciuszko National Park were flooded and the flows of the Snowy River were almost completely diverted, with predictable environmental impacts. Unlike with the Gordon River, no one seemed to mind.

Opening up to competition

By the last decade of the 20th century, the majority of electricity in Australia was supplied by isolated state-owned monopolies. And except for Tasmania, each of these was powered by giant coal-fired power stations, with gas and hydro used to supply additional peaking power (South Australia and Western Australia had higher levels of gas generation). Power prices were low and flat. The abundance of inexpensive, high-quality Australian coal, access to cheaper government capital and the large scale of generation capacity kept Australian electricity prices low in terms of global rankings. Blackouts were infrequent, but not uncommon, and many Australian households had candles stored under the kitchen sink, ready for the occasional outage.

Each utility was run like a government department. There was little innovation and no competition. Every state-owned utility would try, if it could, to over-build capacity in attempts to reduce the risk of blackouts and get ahead of constantly growing demand. Most consumers within each state were charged similar rates, regardless of how much they consumed. Only very large industrial customers, thanks to their direct influence with state governments, could strike big discounts by buying at huge volumes. At its peak, the aluminium sector consumed 14 per cent of Australia's total electricity. Their sheer buying power enabled them to cut deals at extraordinarily low prices. But most other businesses had no such leverage. They just had to take the government price.

By the 1980s, national governments around the world began to realise that their role was changing. As the global economy grew and technologies evolved, governments were struggling to remain competitive suppliers of goods and services. They were still in a lot of businesses: cars, airlines, banks, telephony and electricity, and they found themselves raising increasing levels of debt to fund these subsidised, less-than-efficient enterprises. By the late 20th century, private capital was much more abundant than at the start of the century. The shoe was now on the other foot: governments struggled to sustain the scale of investment needed. In the 1990s, Telstra, the Commonwealth Bank and Qantas were all privatised, along with hundreds of similarly iconic government-owned businesses worldwide (think Renault, British Airways, Air New Zealand, British Leyland). Prime minister Paul Keating's competition policy push brought this wave of microeconomic reform into the state-owned electricity sector.

The way state electricity utilities were run reflected the competing interests of their owners. Governments wanted the increasingly impossible: regular and significant dividends to support state budgets, expansion of the grid to support state development, flat prices, and no blackouts. A risk-averse, departmental culture ensued. In the absence of a market, pricing was arbitrary, set mainly by historical legacy and what voters would tolerate rather than what electricity actually cost. Power stations were lovingly maintained, like an old car in your grandpa's shed. Debts accumulated as the utilities borrowed hundreds of millions for new generation and network assets, but continued to pay government dividends above what they were actually earning. Problems were hidden in balance sheets or state budgets. As the grid got bigger, so did the unsustainability of this model.

A review chaired by Professor Fred Hilmer in 1993 recommended that these regulated, government-owned sectors be opened up to competition in order to drive efficiency and access private capital to finance projects that would help meet growing demand. In response to the review, the utilities were broken up into the four parts of the supply chain: generation, transmission, distribution and retail. The two natural monopolies (transmission and distribution) remained regulated but faced a more rigorous and external review of how much they could charge, while generation and retail were opened to competition. The National Electricity Market (NEM) was established to trade electricity across eastern Australia, with plans for each state (except Western Australia) to be connected by transmission lines, enabling interstate trading and supply of electricity.

The NEM was established in 1998, with three separate agencies created to govern it: a market operator (the Australian Energy Market Operator or AEMO), regulator (the Australian Energy Regulator or AER) and rule-maker (the Australian Energy Market Commission or AEMC). South Australia was connected to Victoria in 2000, Queensland to New South Wales in 2001, Tasmania to Victoria in 2005. (New South Wales and Victoria were already connected through the Snowy hydro scheme before competition policy reforms were introduced.)

The most debt-laden state governments soon cashed in their electricity assets. Jeff Kennett sold Victoria's generators and networks assets in 1995 and used the income from the generous deals struck to pay down debt and build urgently needed public infrastructure. South Australia sold its electricity assets in 1999 to pay down debts. These generators, the retail books, and the poles-and-wires businesses were bought by a combination of Australian energy companies, such as AGL and Origin, and overseas ones, including TXU from the US, International Power from the UK, China Light and Power from Hong Kong and Singapore Power from, well, Singapore.

The big winners were medium- and larger-sized businesses, which discovered they could now exploit their buying power by shopping around between competing electricity retailers. These new savings on electricity made the businesses more competitive. The Australian economy hasn't been in recession since 1992. One of the reasons for this is the microeconomic reforms of the 1990s, and one of those microeconomic reforms was the deregulation of the Australian electricity sector. Competition policy freed up scarce government capital, opened up electricity to competitive markets and reduced costs while enabling a level of innovation and disruption in technologies such as renewables that would not have been possible in a state-owned utility monopoly system.

The new millennium

At the end of 1999, Australians were preoccupied with the threat the Y2K bug posed to the world's computers. They wanted to get a mobile phone and the internet. They were looking forward to the Sydney Olympics and were enjoying new, impressive road infrastructure in Sydney and Melbourne. They felt bad after voting down a republic most of them wanted. Many had a new electricity retailer. Their bills stayed much the same, and most of them weren't thinking much about electricity, energy prices or climate change. But they were about to.

2

HOW DID CLIMATE CHANGE AFFECT ELECTRICITY IN AUSTRALIA?

The politics of electricity in the 21st century is the politics of climate change. And climate change has proven to be a political wrecking ball. While managing climate change has proven challenging to the governments of most developed countries, nowhere has it been as destructive, or as politicised, as in the past 12 years in Australian politics. The career of every Australian prime minister since John Howard has been crippled by climate change, and this seemingly intractable disagreement on climate and energy has now become an issue in its own right.

Understanding the origins and ideology that has produced this rift is crucial to resolving it – ideally, finding a way of returning energy policy to the bipartisan status it held in Australia for most of the 20th century. As we've seen, for much of that century, state governments owned the electricity businesses and their problems.

Demand for electricity grew at the same rate as the economy, and energy was seen as an extension of industry policy. More energy meant more productive workers, more growth and more jobs. Towards the end of the 20th century, tensions arose around stretched state government finances, the resulting creation of a national market, deregulation and privatisation. But this was still a debate about how to get more out of the system: it was about tuning the car, not replacing it.

By contrast, the issue of reducing greenhouse gas emissions is a study in political dissatisfaction. Climate policy is the promise of pain without gain. The best case scenario is that, as a reward for the effort of decarbonising all the world's economies, nothing happens. The worst case is potentially catastrophic. The scale of the threat means that getting high-level support by governments for action on climate change has been easy – initially, anyway. Implementing durable policy measures that deliver these serious emissions cuts has proven much, much harder. Even in economies where there is broad public and bipartisan support for climate change action, this resolve weakens quickly when economic conditions deteriorate. Governments end up trapped between a rock and a hard place, pleasing nobody. Every time they move, they lose.

The science behind the climate

Climate change and the theory that greenhouse gases are driving rapid temperature increases were not designed by radical activists seeking to overturn capitalism and the industrial military complex. It was the conclusion of more than a century of observation and debate about large temperature variations identified over the earth's history, and what caused them. In the early 19th century,

European scientists noticed that giant boulders which belonged high up in mountains had somehow managed to find their way down towards the bottom of river valleys, sometimes hundreds of kilometres from where they originated. They eventually worked out that the rocks had been slowly moved over thousands of years by glaciers that once cut through the valleys of Europe. If there were glaciers once, then why not now? Giant scrape marks through glacial valleys, ice-core samples, deep-sea sediments and tree-ring data revealed that our planet had experienced remarkable periods of change in climate, from eras of tropical heat at the earth's poles to a snow-covered planet. A range of factors had contributed to this variation, including the movement of continents, cycles of solar activity, volcanoes and variations in the earth's orbit. After factoring all these in, the key driver of temperature variation was determined to be the atmospheric concentration of greenhouse gases such as carbon dioxide and methane, which trapped heat in the atmosphere. Higher concentrations of these gases increased temperatures; lower concentrations reduced them. Changes in these atmospheric concentrations over eons were the difference between jungle and glacier.

By the 1970s, a scientific consensus was forming that the industrial release of greenhouse gases, mainly since the start of the 20th century, was driving an accelerated increase in temperatures. (Interestingly, the same theory was also the basis of a confident counter-argument that colder temperatures reported since the 1940s suggested the possible start of a new ice age. This possible threat of a new ice age was the subject of peer-reviewed academic papers and was reported in the mainstream media.) The prospect of global warming began to seep into popular culture. In 1973, Charlton Heston starred in the futuristic dystopian movie *Soylent Green*. It wove the threat of global warming in with the two big pop environment threats of the time: overpopulation and resource depletion. Heston played a detective in a sweaty and overcrowded New York in 2022 who discovers that to overcome a food shortage New Yorkers are being fed reprocessed human corpses.

Backed by better computer modelling capacity in the 1980s, the science firmed around the risk of accelerated warming, like a giant rugby maul finally heading towards one end of the pitch. The moment climate change 'came out' as an issue was on 24 June 1988. On its front page, *The New York Times* published a report on the testimony of NASA scientist Dr James Hansen, who stood before a congressional hearing and warned of dangerous human-induced climate change. In the story, headlined 'Warming Has Begun, Expert Tells Senate', Hansen is quoted as saying it was '99 per cent certain' that the present warming trend was not natural but caused by industrial emissions of carbon dioxide and other greenhouse gases.

This was the second scientific warning about the earth's atmosphere in a decade. A few years earlier, a (different) group of scientists had reported a growing hole in the earth's ozone layer in the upper atmosphere. The ionised oxygen (ozone) that accumulates there protects the earth from most of the sun's harmful ultraviolet radiation, the major cause of sunburn. The ozone was being depleted by the use of chlorofluorocarbon (CFC) propellants in aerosols and refrigerators. As a result of the scientist's alarm, the use of these gases was quickly phased out under an international agreement called the Montreal Protocol, signed in 1987. The global ban was possible because there was a simple and readily available substitute: hydrofluorocarbons (HFC), which had no ozone-depleting properties. The relative cost of switching to HFCs was manageable, with minimal price disruption for consumers.

By contrast, what Hansen and his peers warned of was bigger and harder to manage. To stop the planet getting hotter we would need to rapidly slow and then stop the release of these greenhouse gases into the atmosphere. This wasn't hairspray and deodorant: greenhouse gases were produced by almost everything and everyone. They were emitted every time someone turned on a light, drank a milkshake, built a skyscraper, flew home for Christmas, and bought a coffee, a computer or a newspaper. Despite the magnitude of the scientific warning, the global political vibe at the time was upbeat. Ozone depletion had been swiftly dealt with; how tough could greenhouse gases be?

An Intergovernmental Panel on Climate Change was established in 1988, and in 1990 it first reported the collected views of climate scientists, an earnest re-exposition of the greenhouse gas thesis. It was a time of hope and renewal – the Cold War had ended, the Berlin Wall was down and Nelson Mandela was out of prison. The ghastly pall of nuclear Armageddon had been lifted. An international environmental summit was convened in Rio de Janeiro in 1992, which concluded with the establishment of a United Nations Framework Convention on Climate Change (UNFCCC), a non-binding treaty eventually signed by 197 countries committed to stabilising greenhouse gas emissions. According to the Intergovernmental Panel on Climate Change (IPCC), the biggest global emitters of greenhouse emissions were fossil-fuel (coal and gas) power stations and heating (25 per cent), farming and forestry (24 per cent), industries including steel and cement (21 per cent) and then the fuels from trucks and cars (14 per cent). The frontline sector was electricity, because its emissions were easiest to identify and report. That was good news for countries such as France, Norway and Canada, who got most of their power from zero-emissions nuclear and hydro. It was bad news for Australia's coal-intensive grid.

In the US, the political cards were falling favourably. Republican president George Bush signed the Convention in June 1992. In November that year, Americans elected Democrat Bill Clinton as their new president, with climate change advocate Al Gore as his running mate. With his mandate still white hot, Clinton proposed an economy-wide energy tax early in 1993. But even though the Democrats controlled both houses, Clinton could not get the numbers to support his tax. His fellow Democrats were concerned at the cost and impact of such a tax and wanted instead to focus on economic reform following the 1990 recession. Later that year, Clinton pitched a softer climate change action plan made up of energy-efficiency standards and voluntary programs with industry. The Democrats lost control of Congress in 1994. Since then, national US carbon legislation has never returned to the Hill.

Clinton wasn't alone. The European Commission had been working on its own carbon tax in the lead-up to the UNFCCC meeting. Even with a modest starting price, carve-outs for the poorest EU nations and exemptions for most energy-intense industries, the tax was vetoed. In Australia, the Keating government had ratified the UN deal, and in 1994, then environment minister John Faulkner pitched a modest carbon tax to Cabinet. After months of round-table discussions, Keating couldn't overcome the anxiety about the impact on Australia's carbon-intense energy sector and energy export industries. As in the US, the tax was watered down to a much weaker package of voluntary measures called 'the Greenhouse Challenge'.

There were some successes. Scandinavian countries such as Sweden, Norway and Finland did manage to implement carbon taxes in response to the global deal. Almost all of their electricity came from hydro and nuclear, which have no greenhouse gas emissions. The first Scandinavian carbon taxes operated more like a fuel excise with an environmental intent. What they demonstrated was that the less an economy had to adjust, the easier it was to find a policy response to climate change. This quickly became a major sticking point in getting genuine progress in a global deal to cut emissions.

The Kyoto Protocol

The UNFCCC resulted in an international treaty that became emblematic in the Australian public debate around climate change. It was weaponised politically: if you supported the Kyoto Protocol, you supported action on climate change. If you opposed it, then you opposed action. Both Mark Latham and Kevin Rudd made election promises to ratify Kyoto, which was political shorthand for 'We take climate change seriously, John Howard does not'. When Rudd was elected in 2007 and flew to Bali to make good his Kyoto promise, you could forgive a lot of Australians for thinking that this was the end of the issue: Australia had ratified Kyoto, climate change was fixed.

Except it wasn't. In reality, the Kyoto Protocol highlighted disagreement over global climate change strategy. The central protagonists were the EU, the US and China. At its core was a difficult truth: effective global climate change policy was going to impact economies differently. How do you cut a pain pie? What is fair? Countries with higher emissions – such as Australia – would be impacted more adversely than lower emissions economies. Energy-exporting economies, such as Australia, would be impacted more than energy importing economies, such as Germany. Poorer countries risked being disadvantaged more than wealthier countries, because their ability to develop depended, at least in part, on their ability to access cheap electricity, typically from high-greenhouse-gas-emissions sources such as coal. Besides, their contribution to the problem (the existing stock of greenhouse gas emissions) was marginal.

Kyoto introduced heavy doses of morality to the traditional self-interest of international negotiations. Should high-emissions countries be required to 'pay it forward' for all the 20th-century emissions they had created, even though they were ignorant of the risk at the time? How did the deal reflect serendipity? The emissions from Russia and many former Eastern Bloc countries fell sharply after the end of the Soviet Union, while many European economies had been moving away from coal for economic reasons since the 1980s. Would developing economies be prevented from dragging their people out of poverty? What did this mean for China, which had been developing so rapidly that, by the time Kyoto was signed, it produced the second-highest emissions in the world (now the highest)? Its emissions were growing at such a rate that their continued unfettered increase would overwhelm any reductions made by other economies.

The Kyoto Protocol was championed by the EU, based on its own incrementalist approach to international negotiations: do what it takes to get everyone into the deal, make it stick and then ratchet down emissions in subsequent rounds. The first targets were deliberately soft, only a five per cent cut in global emissions at 1990 levels by 2010. Enforcement was effectively non-existent. It was a soft-start approach: the group representing developing countries (the G77) was offered exemptions from greenhouse gas emissions cuts. Australia asked for, and got, a higher initial target in exchange for its support.

Through its deal-making, the EU brokered broad support for Kyoto. But the US Congress wasn't buying it. President Clinton signed the deal, but only in complete defiance of the US Senate, whose support was required for the deal to be ratified. To be clear, Clinton knew where they stood on the matter: a year earlier they had passed a motion 95–0 opposing ratification of Kyoto without the inclusion of developing countries. This was not a position driven by climate denial, but out of concern for the economic damage the Kyoto model could wreak on the US economy. China was emerging as an economic and global rival to the US, and Kyoto proposed to make the US even less competitive. Democrats and Republicans were in lock-step. No China meant no deal. The US never ratified Kyoto, and the Howard government decided not to ratify in strategic support of the US position.

In the end, Kyoto failed miserably. Global emissions blew out by 35 per cent from 1990 to 2010. Kyoto supporters blamed the US for not ratifying the deal, while in the US, the failure was seen as vindication. But there was bigger fallout. Kyoto facilitated a more aggressive politicisation of climate change in the broader public debate that ensued. US and Australian governments and their supporters were demonised for their moral failure to act on climate change. In response, Kyoto and climate change policy were portrayed as the latest outbreak in an epidemic of political correctness. Kyoto became a metaphor for climate policy more broadly and a reliable rallying point for conservatives. It became a symbol of indulgent, populist action when better solutions existed.

Climate science is complex. So too are the mechanisms to reduce emissions and the physical changes required to transform complex industries such as electricity, steel and agriculture. Perceptions matter. In the end, it is what most people go on. It probably didn't help the credibility of the Kyoto argument that most European countries didn't always back their moral rhetoric with action. In 2003, the EU agreed to implement their own emissions trading scheme. Because the legal operation of this lay at the national level, each member country was asked to submit their starting allocations of permits on the exact same day in 2004. In other words, they each had to nominate how much they would constrain the emissions of their power stations and factories without knowing by how much their fellow member countries would cut theirs. It was, without equal, the biggest international game of chicken ever.

When the numbers were revealed, most EU member countries had set constraints larger than their greenhouse gas emissions at the time, some by as much as 25 per cent. Only the Germans and Slovenians cut their emissions as intended. The European Commission worked hard to get the scheme back on track, and by 2008 carbon emissions were trading as intended. But the reputational damage was done. The EU liked to talk a big game on climate, but when it came time for their member countries to deliver, they ratted on the deal.

The climate changes: drought

When the Liberal–National Coalition was elected in 1996, climate change was not on the political radar of most Australians. John Howard's political brand was stability, and his method was to listen to what the majority of the electorate wanted, and then play it back to them. On climate change, Howard was risk averse. Australia continued to participate in global climate negotiations, more to keep an eye on them than lead them. After gun-law reform following the Port Arthur Massacre in 1996, Howard's main focus was the introduction of a goods and services tax. In the 1998 election contested by Labor's Kim Beazley, climate change was not on either major party's political radar. Nor did it get a mention in 2001.

However, Howard did not completely ignore climate. On the urging of Environment Minister Robert Hill, he created the Australian Greenhouse Office in 1998 and introduced the modest Mandatory Renewable Energy Target scheme (MRET) in 2001. His own former Resources and Energy Minister, Senator Warwick Parer, recommended a national emissions trading scheme in 2002 as part of an independent review of energy markets. A plan to implement emissions trading was presented to federal Cabinet a year later, but Howard vetoed it for being too expensive. Still, as part of negotiations with the Australian Democrats on the GST, Howard agreed to introduce a fledgling household rooftop solar photovoltaic (PV) scheme. Within a decade, this seemingly trivial scheme would help influence the shape of the electricity grid in Australia.

It wasn't until 2004 that a Labor leader actually mentioned climate policy in a federal election campaign. Opposition leader Mark Latham called for the ratification of Kyoto, a carbon trading scheme and increasing the MRET from two to five per cent. The issues were not prominent in the campaign. Howard was elected with an increased majority and control of the Senate. Labor's interest in climate had more to do with the rise of the Greens, who at the 2004 election effectively replaced the Australian Democrats as the third party in Australia, and started to cannibalise Labor's primary vote in the inner city.

The defining moment for climate change policy in Australia was the Millennium drought. By 2006, this drought – the worst on record – was at its apogee. Reservoir levels in Melbourne had fallen to 38 per cent, and in Sydney to 33 per cent. In November that year, watering lawns and filling swimming pools in Australia's two largest cities was banned. Other major cities had similar restrictions. The gardens of Australia's relaxed, sleepy outer suburbs were dying, their backyard pools taken out of commission. A primarily suburban nation that had been vaguely disinterested in climate change suddenly thought it had arrived, almost overnight. And they were terrified.

Al Gore, unsuccessful climate politician, had evolved into Al Gore, climate entertainer. *An Inconvenient Truth* may not have had Australians rushing to see it, but its release did coincide with the crushing sense of dread that consumed Australian voters. Howard didn't see it coming. In a major speech in July, he outlined his vision of Australia as an energy superpower. In true Howard style, he mentioned everything: coal, gas, renewables, oil and nuclear. But by Christmas that year, his vision seemed suddenly anachronistic. Australia's rejection of the Kyoto Protocol, previously politically benign, was now cast as the cause of the suburban drought. Howard's popular support started to collapse. As water restrictions were dialling up to 11, he accepted an invitation to speak at the Business Council of Australia's annual dinner in Sydney. That night Howard was hosted by the Council's President Michael Chaney, a pillar of Australian business elite and petroleum geologist with a blue-chip corporate CV. His brother Fred was even a deputy leader of the Liberal Party. This was, traditionally, the safest of safe rooms for a Liberal prime minister.

Then, Chaney dropped a bombshell. The Business Council of Australia announced it was backing decisive action on climate change – more specifically, a national emissions trading scheme.

Chaney's speech was more than just helpful advice. What he said to the Prime Minister was that business had moved decisively on climate policy. It was a risk that needed to be actively managed. 'I must say I am of the school that thinks of this issue in the same way that I think of home insurance: I doubt if my house will burn down, but I'm prepared to pay a premium just in case.'

Three weeks later, Kevin Rudd deposed Kim Beazley as leader of the opposition. Smelling the blood in the water, but, like most Australians, with little understanding of the issue itself, Rudd advised he would be back in the new year with an aggressive climate reform package. He hired an environmental policy expert to sit in a dark room for three months and write him a climate policy blueprint, which was released in March 2007. It had all the toys: emissions trading, increased renewables, clean coal, ratification of Kyoto, energy efficiency and pink batts. Rudd did what no Labor leader had done before him: put climate at the top of the political agenda.

Like an aircraft carrier under fire, Howard wheeled for all his life on climate. A week after Labor's leadership coup, he announced he would investigate an emissions trading scheme. The Shergold report was released in July 2007, three months after Kevin Rudd's 'cram for the exam' blueprint. The review driven by senior bureaucrat Martin Parkinson went further than Howard had intended, but by this stage it was largely academic. The voters had stopped listening. The Coalition was comprehensively defeated at the November 2007 election. After the dust settled, Rudd and Penny Wong, Minister for Climate Change and Water, headed off to the international climate change negotiations in Bali, where, to a standing ovation, they committed to ratify the (now mostly redundant) Kyoto Protocol. It was a short-lived victory. Rudd was soon to discover the political death trap that was climate policy.

The Howard–Rudd skirmish on climate inadvertently established the rules of political engagement on energy policy for the next decade. Major economic policy, like a national emissions trading scheme, was an off-the-shelf announcement. Its design and implementation remained subject to detailed consultation and review. The real political differentiators were the other policy toys in the show bag. Election promises such as a '20 per cent by 2020' Renewable Energy Target (RET; a revised set of targets to replace the MRET) were developed in confidence and announced to maximise political impact, rather than minimise economic or technical harm. No one was consulted about the broader impacts of the RET until after it was announced as an election promise and its implementation was being finalised. Many of these subsequent market interventions were far more consequential than their sponsoring parties realised at the time.

This precedent of surprise announcement by governments on electricity market interventions has continued ever since. The process of consultation in designing policy that impacts complex, economy-wide systems like energy markets is not some sort of political correctness. Most of the intellectual horsepower exists outside of governments, among the hive mind of energy companies, networks and transmission businesses, governance agencies, consultants and academics – all of whom live and work around the machine every day. It says a lot about how ideological and political energy policy has become that, as the system has descended into greater dysfunction, governments have felt less obliged to check with the experts before intervening.

Crisis point

Australian climate politics in the mid-noughties was like an episode of *Top Gear*. Howard drove climate understeer; Rudd drove climate oversteer. KRudd's political brand was action with urgency. He was determined to differentiate himself completely from Howard on the issue that helped him into the Lodge. What he lacked in experience on energy, he tried to make up for with enthusiasm. Following his trip to Bali, with the applause of the world still ringing in his ears, Rudd set up a new climate change department and charged it with building him an emissions trading scheme. It was the policy equivalent of the Apollo space program. The new department had until July 2008 to draft an 'introductory' green paper for Rudd's Carbon Pollution and Reduction Scheme (CPRS), which scoped out some of the possible design elements. It was more than 500 pages long.

The frenetic pace to create this complex emissions trading scheme was made easier with some cut-and-pasting from the EU design, which had been legislated two years earlier. Additionally, wounded by Rudd's sweeping electoral success, the Coalition put up little opposition. Business had signalled it would back a well-designed scheme. Rudd was further encouraged by growing optimism about a breakthrough international climate deal to be ratified at the Copenhagen climate change conference at the end of 2009. Barack Obama became the Democrat candidate for the November 2008 presidential election, campaigning on an 80 per cent emissions reduction target by 2050. Malcolm Turnbull, a supporter of emissions trading, became Opposition leader on 16 September 2008. It was probably the day that Rudd, and the world, got closest to a global climate deal.

Because the day before, the world had changed. Lehmann Brothers had collapsed, triggering the global financial crisis.

Climate policy momentum evaporated, and Rudd's political focus immediately shifted to avoiding recession in Australia. The ensuing global recession both stopped progress on new policy and unwound existing schemes. The EU effectively parked its new emissions trading scheme by letting the price of carbon emission fall to nearly zero. Before Obama could even get in the White House, his ability to shift the US towards a more effective climate agenda was fatally wounded. The climate summit in Copenhagen at the end of 2009 went from a party to a wake.

Rudd avoided recession in Australia and persisted with his emissions trading bill. Evaporating global action on climate left Australia isolated and the government's political mandate weakened. Affluent, inner-urban voters placed a higher value on future environmental outcomes and wanted Rudd to continue, but more price-sensitive suburban voters grew anxious about their economic security and became reluctant about anything that imperilled this. Still, climate policy's next victim wasn't Kevin Rudd – it was Malcolm Turnbull. Turnbull wanted to negotiate some reforms to the government's scheme, but support it. Conservatives saw no reason for Australia to persist with emissions trading until a global agreement emerged. This galvanised support against the scheme and Turnbull. The CPRS eventually got knocked out twice in the Senate in 2009. Tony Abbott defeated Turnbull in a leadership challenge at the end of 2009, just as it started raining again. The drought broke. The moment for a bipartisan national emissions trading scheme was over.

Rudd's demise followed shortly after. His eventual admission that he would delay emissions trading dismayed the left. His 'action with urgency' brand now looked like over-reach. Backing down simply played into the Coalition's hands and emboldened Abbott and his conservatives, who had argued all along that the approach was reckless and unfounded. Rudd's successor Julia Gillard was later forced back to reprosecute a hybrid carbon tax and emissions trading scheme after having to form a minority government in 2010 with the Greens. By then the damage had been done.

The drought was over and the world kicked climate change into the 'next year's problems' basket. Abbott campaigned openly and aggressively against Gillard's carbon tax ('axe the tax') to win the 2013 election and then sought to erase all support of emissions trading or pricing carbon from the political landscape. His solution was the ironically named 'direct action', which returned Australia to the mid-1990s, involving little more than the pantomime of reducing greenhouse gas emissions without imposing any material costs on the economy. It was a fragile calm.

The business risk

The role of business has been important in the evolving attitudes to climate change. This is because in an apparently moral issue, business is amoral. Its ruthless decision-making and cold, calculating self-interest make it a relatively stable and predictable agent in the debate. Business is a necessary ally and a difficult opponent. The personality of business was usefully summarised by Canadian lawyer Joel Bakan in his 2004 book *The Corporation*. In it, Bakan likened the behaviour of large companies to that of a psychopath. The single unambiguous purpose of corporations is to maximise returns to shareholders. This is an amoral purpose. Each company will do what it needs to do lawfully to protect and enhance its ability to deliver shareholder benefit.

Mostly this is a net positive. Profitable companies are constantly improving efficiency and seeking to innovate. A corporation will actively work to protect the safety of its employees both because it is required to do so by law and because a safe workplace that values employee welfare is more productive. The recent emergence of 'corporate activism' is simply an extension of this amorality. Companies sometimes take positions on social or other issues that are not directly relevant to their business, such as the approximately 1000 companies who supported the same-sex marriage referendum. This made sense: it reflected the values of the majority of their employees and customers. Corporations will selectively take a position on a public debate where they are either defending their business interests or enhancing the company's brand. The action they take is no less authentic, but it is always strategic.

When climate change first emerged as an issue, the reactions from businesses were mixed. Those most affected by it, such as energy and resources companies, paid more attention than others. At first they tried to find ways of avoiding being impacted by any policy measures or delaying their implementation by warning of the economic impacts of such measures. In some cases, they emphasised the scale and complexity of the science and questioned whether such consequential action should be taken before there was greater certainty. This has been described as a climate change conspiracy, but in reality this is precisely what you would expect these sorts of businesses to do in the circumstances. Governments receive a constant stream of self-interested lobbying from companies and industries, as well as from other sectors of society. It's how they respond to it that counts.

As the debate continued, growing interest was shown by a wider range of businesses who would be less adversely affected by emissions-reductions policies. Computer and IT companies such as Apple and Microsoft began taking measures to reduce their greenhouse footprint, in large part because it reflected the future-looking nature of their companies to an engaged part of their customer market. By the start of the 21st century, climate public relations emerged as specific businesses announced steps to go carbon neutral or install renewables. Again, it was precisely what you would expect them to do.

There was one business sector that had been paying particularly close attention to climate change: reinsurance. These global businesses underwrote the risks for domestic insurers and were constantly tracking the risks of everything they insured, which included the damage from storms and other weather events. By the early 21st century, the data started to show that both the size and frequency of storm and flood events was increasing. The rate of payouts was growing much faster than the natural increase in the number and value of buildings and other assets damaged by such storms. This trend started to emerge in the early 1990s with Hurricane Andrew and continued to Hurricane Katrina in 2005. Reinsurers had no reason to talk up the problem. It caused them to recalibrate climate change not as a science and policy issue, but as a risk issue.

In the absence of binding laws on greenhouse gas emissions, many businesses struggled to get their amoral heads around climate change. But they all understood risk. As a risk-management strategy, climate policy made complete sense. The science no longer had to prove beyond doubt, it simply had to prove sufficient risk to warrant policy action. Businesses realised that in addition to the reputational risk of investing in greenhouse-intense assets, they faced policy risk from potential impairment of assets in the future by climate-related laws and even liability risk from future class actions. The shift in the cost of climate risk from reinsurance flowed through to financing and banking. This made it increasingly difficult to find competitive finance for carbon-intense long-life assets, such as coal-fired power stations. Separate from this, the banking sector also realised that if there was some kind of new market for carbon trading, it would create a multi-million-dollar service industry to process the transactions. Banks liked the idea of carbon trading.

These risks are no longer theoretical. In January 2019, California's largest energy utility, Pacific Gas and Electricity (PG&E), filed for bankruptcy. *The Wall Street Journal* described this as 'the first climate-change bankruptcy'. The company had been exposed to around $30 billion in liabilities and 750 lawsuits resulting from bushfires that were linked to its network of power lines. Increasing heat and dryer weather in California escalated both the risk and the scale of potential liabilities for the utility, which, ironically, has been pro-active in reducing its greenhouse emissions.

Climate unease

The science that underpins climate change is complex. It is multi-disciplinary, incorporating input from multiple fields of science: climatology and meteorology, atmospheric physics and chemistry, geology, oceanography, biology and paleobiology, meteorology and astronomy. The mitigation of greenhouse gases is also challenging. It requires sophisticated policy design to efficiently guide investment and encourage technology development. It will require major technical changes in many industrial and agricultural processes. It will change the way people live, how they move and even what they eat.

Mitigating greenhouse gases involves decades of sustained cost, change and constraint, so getting broad-based support for the policies chosen and the need to act is critical. This requires a leap of faith from the public. They need to trust the advice they are getting before they'll have the resolve to continue with a generational process of sometimes painful structural, economic and social reform. Achieving that is not easy, particularly given it needs to be agreed and enacted at a global level to work. And as with all science, there is always some shade of doubt.

Attitudes to climate change tend to be cast as a binary state: you either accept the science, or you do not. There's a third state, which I call climate unease. Climate unease reflects a centrist view: a rational respect for the advice of expert scientists undermined by the sometimes hysterical, inflammatory and inaccurate nature of the claims made by some more vocal advocates. As the debate on climate escalated rapidly towards a political and economic decision, those who were previously disinterested found themselves faced with a decision. What do I believe? Who do I trust? In the mid-2000s, this was a majority of Australians: business leaders, commentators and ordinary punters. The complexity of the information on which such views were based meant that these personal opinions were guided by influencers and leaders at the time.

In 2005, Australian paleontologist Tim Flannery published his landmark book *The Weather Makers*, breaking down the science of climate change for a lay audience. It was a global bestseller. Flannery is a scientist but not a climatologist. Nonetheless he became a leading public face of climate science as, along with an emerging group of climate scientists, he sought to explain and advocate for action on global warming. In 2007, he was awarded Australian of the Year and was later appointed Chief Commander of the Climate Commission by the Gillard government.

As the debate escalated, Flannery made some bold claims to try to illustrate the future risk of climate change. He suggested that Adelaide and Brisbane may run out of water by 2009, Sydney by 2007. In 2004, he said there was a fair chance Perth could be the first ghost metropolis of the 21st century. He wasn't alone. In a similar vein, Al Gore's movie *An Inconvenient Truth* stretched the timeframes of some of the potential consequences of climate change for dramatic intent: like predicting there would be no snow on Mt Kilimanjaro by 2015 (there still is). Climate scientists started to air their opinions on climate policy design, an area in which they held no expertise. At the time, these and other claims were part of a much broader emerging debate on climate, but they have been replayed by climate sceptics ever since. While such comments probably convinced some Australians of the need to address climate change, they hardened opposition as well.

Thirty years since James Hansen warned that the planet was warming, there has been no great reveal in climate science to suggest the core thesis is wrong. Anyone is entitled to disagree with climate science, and we can expect that some will continue to do so. It's also evidently lucrative for some to reject the advice of genuine experts. Most remaining high-profile climate sceptics have little idea about the details of the science they claim to doubt, they just 'feel' that there is something wrong with this debate. They should be free to express their views. But this should not distract the discussion about what we do next. Overreacting and being wrong about climate change is a much better outcome than doing nothing and being right.

Politicians of all colours have demonstrated remarkable flexibility on climate policy over the past decade. Prominent climate sceptics have backed decisive climate policy; supporters have deferred or even opposed action. There remains a dogged reticence by a section of society to treat climate change as a credible risk. This climate reluctance has endured in the political world as it shrinks in the real one. It acts like a smouldering ember in a dying fire that relights in the slightest breeze, and it recharges delay and doubt. Except governments are discovering that, at least when it comes to frontline sectors such as electricity, ignoring the problem has become the problem. In the absence of political leadership, global finance is driving the future of climate policy – not out of some moral need to lead, of course, but because it has risks to manage.

So if we can't save the planet, can we at least save the electricity system?

3

WHAT CAUSED THE ELECTRICITY CRISIS?

The 20th-century electricity industry was brutalist by design, gigantic in scale, judicious in its investment. The coal-fired power stations built in the second half of the 20th century were up to half a kilometre long, took seven years to build and ran for fifty. Transmission lines pumped large volumes of electricity for hundreds of kilometres which then fed nearly a million kilometres of power lines running down city streets and country towns. The system was built to endure. Given the monolithic time scale of building new capacity, it was prudent to over-engineer. Electricity grids ran a deep reserves bench of generators to cover all contingencies.

When climate and energy policy exploded into the Australian political zeitgeist in 2006, the operation of the electricity grid was initially unaffected – the grid's biggest concern during the Millennium drought was finding enough water for the hydro and some of the coal generators (fresh water is required by some coal generators for cooling). As electricity demand was still constantly increasing with growth in the economy, an informal schedule of planned and proposed power stations were queued up and ready to start building when needed. It was all comfortingly dull.

As the political fight on climate and energy raged between John Howard and Kevin Rudd, it seemed possible that the debate might revert to a unified outcome. Both major parties took basically similar designs for emissions trading schemes to the 2007 election. As the implementation of comprehensive, well designed and national climate policy broke down, so too did any hope of unity. When the lead agent of reform became the renewable energy targets, the political debate re-formed around the new policy.

The public debate on energy policy polarised into a cross-town derby between the Coal Powered Rationalists and the Renewable Energy Reformers. Conservative versus progressive. Each side celebrated the importance of their favoured technology, while mocking the failings of their rivals. This name-calling and baiting became standard in public commentary and on social media. The feud largely ignored the important but different roles these technology types played in the operation of the grid and during heat waves, and their influence on power prices and emissions. Extreme, hardcore supporters on opposite sides talked up new coal-fired power stations or confidently predicted the imminent arrival of a 100-per-cent renewables grid, even though neither of these options was seriously plausible. This rivalry continues to this day.

Carbon price gridlock

In 2007, the headline political debate in Canberra was the sudden race to see who would be the first to introduce a carbon price. Underneath the waterline, a quieter but equally important change was taking place in global capital markets. Investment banks and other sources of funds used to build multi-billion-dollar infrastructure (such as power stations) were changing their lending rules to reflect the emerging carbon risk. One of the most impacted forms of infrastructure was the coal-fired power station: these were long-life assets that were expensive to build and had high levels of greenhouse gas emissions. In the 1960s and '70s, coal-fired power stations were safe as houses, the type of asset built by pension funds and others seeking reliable, low-risk returns.

Early in the 21st century, things had changed. It was now much harder to see how and if new coal-fired power stations in Australia could remain profitable ten years into the future, let alone twenty. Until there was greater clarity and stability around how this carbon risk would be monetised, the chequebook would remain closed. New coal could still find backers in some other parts of the world – particularly in emerging developing countries, where the asset was backed or underwritten by the national government and/or a major institutional financier such as the World Bank. In Australia, though, new coal-fired power quietly became unbankable.

But the grid was still going to need new power stations. So if coal could not replace coal, then something else had to. Without much fanfare, the informal schedule of new power station projects quietly shifted from coal to gas, a less carbon-risky type of generator. Gas had half the emissions of coal, the power stations were cheaper to build and they had a shorter life to recoup their costs. But the electricity was more expensive to produce. In 2007 renewables had no emissions, but were more expensive again.

The intent of an emissions trading scheme or a carbon tax was largely the same. They involved a government putting a monetary value on the amount of greenhouse gases emitted. This made lower-emissions generation cheaper and higher-emissions generation more expensive. Emissions trading did this by setting a specific limit on the amount of greenhouse gases permitted each year and then letting liable parties trade these permits in a market, thereby setting the price. The scarcer the permits, the higher the price. Under a carbon tax this price was pre-set by the government. A starting carbon price of around $30 per tonne of carbon dioxide was considered enough to make lower-emissions gas-fired generation competitive with coal.

The economic theory on pricing emissions to transform the electricity grid is not unlike a game of musical chairs: you start with the established amount of existing greenhouse gas emissions from power stations and other major point sources (the chairs). Then you allocate permits to emit these gases, but make the allocation lower than what is currently being produced (i.e. remove a chair). The next year you lower it again (remove another chair), and again the following year (and another). As you slowly and transparently ratchet down the number of emissions permits, they become scarcer and more valuable. Existing and new entrants have an increasing financial incentive to invest in new ways of producing the same amount of electricity with fewer emissions. The least efficient, highest-emitting power stations are most likely to close first. The cheapest, most efficient sources of lower-emissions generation will replace them.

Whether a government chooses to trade emissions, tax them or use some other related mechanism, the most important ingredient for a climate policy is confidence. Whatever is chosen, this type of system will need to endure for decades. Investors will need to base decisions they make today on knowing how the scheme will work years into the future. If there was a risk that the scheme might close or be dialled back, then owners of high-emissions assets are likely to wait before exiting the market, hoping to extend the life of their assets. And investors in replacement technologies would also wait, as the conditions for their new assets to be profitable depend on the scheme rolling out and the old generators exiting. When there is no confidence, everyone stops. It's gridlock.

This is precisely what happened in Australia in 2008, after the global financial crisis and the subsequent moonwalk away from carbon pricing in 2009. When Julia Gillard replaced Kevin Rudd in 2010, she had no plans for a carbon price. Gillard's Clean Energy Act (three years of a carbon tax followed by emissions trading) was only introduced in 2012 after Labor was forced to form a minority government with the Greens. With Opposition leader Tony Abbott vowing to repeal it and looking likely to win the following election, the tax introduced did nothing to change or stimulate new investment.

Emissions did fall by about five per cent during the two effective years of this short-lived carbon tax, but this was largely due to how the policy changed the generation strategy of two businesses: Snowy Hydro and Hydro Tasmania. The carbon tax pushed up the wholesale price of electricity as coal generators tried to pass on the cost of the new carbon taxes. Hydro generators got paid the revenues from the price increase, but paid no carbon tax. Knowing this would occur, hydro generators withheld as much water as possible before the tax started and then increased their output during the two-year carbon tax period. This net increase in zero-emission generation reduced total emissions for the period of the tax. But it was unsustainable simply because they would have run out of spare water.

Moving targets

Kevin Rudd went to the 2007 election armed with his toy box of climate change policies. His political sweetener to emissions trading was an increased Renewable Energy Target (RET). In 2001, during the last years of a bipartisan energy position, John Howard had introduced an entry-level two-per-cent RET. Mark Latham had committed to increasing this to five per cent in 2004. Rudd's election strategy was as simple as it was brutal: to keep his foot on Howard's throat. The design principles for Rudd's expanded RET could be described in just two words: repetition and differentiation. Rudd promised a legislated target of 20 per cent of all of Australia's electricity generation by 2020. Bigger was better.

Renewables policy was first developed not to combat climate change, but in response to the oil crisis in the 1970s. US President Jimmy Carter offered tax credits to wind energy in 1978 as one of a number of measures to develop increased energy security. Energy security was also a key driver of renewables schemes in Germany and other European countries, as increased wind and solar reduced their uneasy reliance on importing gas from Russia. Renewable energy soon developed a cool brand: futuristic, smart, clean and local. It became the carrot of climate policy to the stick of carbon pricing. After President Clinton failed to get a modest carbon tax through in 1993, he did manage to get modest support from Congress for renewables tax credits. Renewables policy more easily crossed party lines. President George W Bush set renewables targets in his home state of Texas and even put solar panels on the roof of the White House in 2003.

In 2009, after the GFC, Kevin Rudd's headline policy focus shifted. KRudd's election-mandated climate-change package had been replaced by KRudd's $42 billion-worth of pop-up government spending programs, designed to help stave off recession in Australia. Most of the money went into school halls and cash handouts. In energy there was also some sugar: bigger subsidies for rooftop solar hot-water systems and the (now infamous) pink batts scheme, which gave away free ceiling insulation to some of the 2.7 million Australian homes still without it.

For a while there it looked like the RET would go down with the emissions trading ship. Senior bureaucrats were privately dismissive of the RET as populist and poorly designed. What modelling was done focused only on cost, and there was no consideration of the potential technical impacts and risks.

Politics is a combination of many elements: ideology, good intentions and opportunism among them. It was a lot easier to sell a popular new technology subsidy than an unpopular power-price increase. A curious alliance between Climate Change and Water Minister Penny Wong and Shadow Energy and Resources Minister Ian MacFarlane saw the RET cut free from the sinking wreck of emissions trading, and it was legislated in September 2009 with bipartisan support (the Greens voted against it). At the time, it was assumed the RET would mostly finance wind farms and the emerging geothermal projects (solar PV was still considered a minor player). The new renewables were expected to have little impact on the role of the existing generators, as they were expected to supply the expanding demand for electricity – a demand that had been reliably predicted since the 1950s. What could possibly go wrong?

By the time Kevin Rudd flew back from the climate negotiations in Copenhagen at the end of 2009, his hastily assembled pre-election climate policy had been completely remodelled. His Emissions Trading Scheme was in the political bin. An unplanned and populist renewables target had been inadvertently recast as the new centrepiece of a heavily compromised climate strategy. The survivors were a mish mash of press-release policies: subsidies for the equivalent of around 5000 new wind turbines over the next 11 years, free ceiling insulation, subsidies for solar hot water and rooftop solar PV. Electricity businesses were left scratching their heads. They couldn't build gas generation because it was too expensive and they couldn't build coal because it was unbankable. So they started building wind farms because, well, they had to.

South Australia loads up on renewables

Pretty quickly, Rudd's patched-up climate plan started to veer off script. The cost of rooftop solar PV started to fall rapidly, and the systems were selling like hot cakes. This was great for households (and important strategically), but it meant the budgeted cost of solar subsidies blew out fast. The well-intentioned but poorly executed free home insulation scheme went off the rails; making insulation free attracted a cowboy element to an inherently risky installation process, and four installers were tragically killed. The rushed design of the hot-water subsidy also resulted in design flaws. Some installers realised they could get the price of new hot water down to nothing by massively over-engineering heat pump installations. Large banks of heat-pump systems were turning up in odd places, such as football change rooms. The geothermal industry, which had been expected to do much of the heavy lifting in supplying renewable energy, instead went broke.

There was another surprise. Since the end of the Second World War, electricity demand had increased in lock step with economic growth. This made sense. Population increase boosted electricity demand. Also, labour productivity grew as people used more electrically powered machinery, which contributed to rising wages. In the last great recession of 1991, electricity demand dipped when the economy shrank.

This seemingly iron-clad economic relationship ended in 2009, just as the ink was drying on the RET legislation. The Australian economy kept growing, but demand for electricity started to fall. It kept falling until 2014 and has hovered uncertainly ever since. This decline in demand has been attributed to a range of factors: the shift in the Australian economy from high-electricity-consuming industries (aluminium and cars) to lower-consuming services (tourism and education); the impact of efficiency improvements in appliances, machinery and dwellings; consumers reducing demand in response to higher electricity prices; and the increased uptake of rooftop solar PV and solar hot water.

What this meant for the grid was that the new renewables started to displace existing generators. From an emissions perspective this was a good thing, but the potential consequences hadn't been seriously thought through. No one knew if demand would keep falling or return to growth. No one thought through which regions the renewables would be installed in. No one considered how the existing generators would respond. So the electricity grid lumbered on, wind farm salespeople rubbed their hands with glee, and solar installers started running TV ads on primetime television and sponsoring football teams.

It was never particularly clear what the objective of the expanded RET was. It was part industry policy, part emissions reduction scheme and part populist subsidy. One view was that the RET would force the electricity sector to discover how renewables worked at scale. The target did reduce emissions: every megawatt hour that a wind turbine or solar panel generated saved up to a tonne of carbon dioxide. It was definitely a subsidy: the scheme funded the gap between the (higher) cost of renewables and the wholesale price set by coal or gas. To keep this cost as low as possible for consumers (who paid for it through a four per cent increase in their power bills), the target was designed to subsidise only the cheapest renewables.

This had two consequences. First, it meant most of the large-scale renewables generation used the cheapest technology: wind turbines. Second, it meant that these wind farms were installed where they earned the best returns. South Australia delivered better returns than other states because it had historically higher wholesale electricity prices, its State Government made it easier to develop wind farms and, of course, they had a lot of good wind. By 2016, according to Electricity Gas Australia, most of the renewables built under the RET were wind farms, and 42 per cent of them were built in South Australia.

At the same time, rooftop solar PV was going gangbusters. The combination of generous state and federal subsidies and falling costs saw installations skyrocket after 2010. South Australia was per capita the biggest rooftop solar PV state in the country, ahead of Queensland and Western Australia. By 2016, solar was on more than 30 per cent of all dwellings in South Australia and Queensland. In some Adelaide suburbs up to 70 per cent of households had solar PV. Household solar systems were much larger in number but smaller in total generation than wind farms. By 2016, around seven per cent of South Australia's total electricity supply came from solar PV, while 34 per cent came from wind.

As the share of wind and solar increased, they began to dramatically change the way the grid operated. Rooftop solar

PV was more predictable – it worked like many of us: slow in the morning, flat-out at midday, cooked by sunset, rubbish on cloudy days. Wind was less predictable, resulting in increasing, semi-random swings of wind-generated electricity that vaguely followed seasonal weather patterns. Sometimes the wind would blow hard on high-demand hot days, sometimes it wouldn't. In winter and spring it would blow in the middle of the night. In short, the amount of electricity generated by wind farms was entirely unconnected to the needs of the grid.

Expanding renewables generation required new ways of thinking about electricity generation. 'Baseload' power refers to technologies such as coal and nuclear generation – technologies that operate like a giant combustion heater. They take a while to start up, and once operating, they need to stay operating, dialling output up and down within a limited range. The use of renewables at larger scales focused attention on the need for firm – as discrete from baseload – generation. 'Firm' generation describes bulk electricity supplies that can be controlled: switched on and off and dialled up and down to balance the constant variances in solar and wind output. Baseload and firm are related, but not the same. At lower levels of renewables generation, big generators such as coal-fired power stations were able to 'firm' variances in renewables (by slowing or increasing their output). So too could technologies such as hydro and gas – and, in specific, short-run situations, batteries. As renewables expanded, they generated at such scale that the balancing requirements fell outside the operating range of coal-fired generators. Big renewables needed to be firmed, but this was increasingly unsuited to coal baseload generators.

As the volume of renewables started to increase, the coal and gas generators in South Australia were adjusting their generation around the weather. Electricity demand had always been a weather-related business, but now there was an extra dimension. When the wind dropped, generators would ramp up output. When it blew, they would dial right back. Power companies were learning, in real time, about life with lots of wind energy. Eventually, there were times when the wind generation at 4am was bigger than the entire demand in South Australia. The surplus was sent into Victoria, where it wasn't really needed. By 2016 South Australia had, by accident, become a world leader in renewable energy integration. The system engineers still managed to operate the grid reliably, and the RET continued to mandate more wind farms. Rooftop solar kept selling and selling. In 2013, renewables made up 30 per cent of South Australia's total generation. Three years later they exceeded 40 per cent.

Given the popularity of renewable energy, the growth of renewables generation in South Australia was hailed as a success story. The State Government took credit as a pioneer of renewables integration. The new wind farms created an oversupply of generation that pushed down wholesale electricity prices, a fact that was spruiked as an even better consumer story. Clean, green and cheaper for everyone! Subsidising more expensive generation technologies lowered energy prices! It was an impossible win–win. For a while, the South Australian renewables story was a political utopia for then premier Jay Weatherill: a case of premature self-congratulation.

However, below the political waterline, things were starting to get dark both commercially and operationally. The South Australian electricity grid was the smallest of all the mainland grids, connected to Victoria by two transmission lines known as interconnectors. These lines could supply about 25 per cent of South Australia's maximum peak load, pumping in cheap brown-coal power from Victoria's Latrobe Valley. To help back up the increased number of renewables, the Australian Energy Market Operator (AEMO) agreed to a 40 per cent increase in the size of the larger Heywood Interconnector. Otherwise there were three main power stations in South Australia: the Northern Power Station at Port Augusta, which was the last remaining brown-coal plant in South Australia, and two large gas generators at Torrens Island and Pelican Point. There was also a scattering of smaller peaking power stations designed to jump in to help when demand and prices were high.

The big swings in renewable energy started to pose major technical problems for the Northern Power Station. As wind generation increased, Northern was more often generating to lose money, and being forced out of the market altogether on windy nights. Something had to give.

In 2015, Northern's owners, Alinta Energy, tried turning it off over winter (except in emergencies) and running it only during summer. They tried contracting industrial customers, hoping to shore up enough regular demand to keep the power station operating. They even asked the State Government for $24 million to tide them over for another three years. No one was interested. Wholesale prices were falling. Who would want to get in the way of that? Governments pointed to the rise of renewables and falling prices as clear evidence that their interventions were working. They had mastered the giant NEM machine and made it do their bidding. And so Northern closed in May 2016. It was the second-biggest power station in the state.

But the problems didn't stop there. South Australia's two main gas generators were slightly more flexible and could cope better with the wind variations, but the top-up power from Victoria tended to get the lion's share of the work as it was even more flexible, and cheaper. Pelican Point, the more expensive of the two gas power stations, was soon more often off than on, and its owners started heading down the same path as Northern. Half of its capacity was mothballed in 2015. The electricity market had not been designed to cope with this. Intermittent renewables generation was being forced in at scale by mechanisms completely outside the design of the market, and they were, in turn, wiping out the firm generators needed to back them up. The South Australian grid was becoming more fragile.

At the start of 2016, the then boss of the AEMO – which is like the air traffic controller of the grid – was a guy called Matt Zema. Zema, a South Australian, had become increasingly anxious about conditions in his home state. He could see the fragility of the grid and yet he was powerless to act, because nothing had happened except prices were down and a coal-fired generator had closed. Like Roy Scheider at the start of *Jaws*, Zema knew something was wrong, but he had no justification to close the beaches. When Northern closed, South Australia was in uncharted water. The big gap in firm (dispatchable, controllable) supply sent wholesale electricity prices up by 23 per cent, and power bills with them.

Six weeks after Northern closed, a winter cold snap hit South Australia. In the second week of July, electricity demand increased sharply as Adelaidians put on their electric heaters and reverse-cycle air conditioners. Then the wind stopped blowing. The state's gas generators struggled to source enough gas to operate at capacity. The interconnector to Victoria was partway through its upgrade, leaving its capacity constrained. Wholesale electricity prices skyrocketed. That was the first serious market warning that the state was running out of power.

Matt Zema tragically and suddenly died later that month. It was like losing the captain of a ship just as it was about to navigate particularly hazardous waters.

System black

In the end, the trigger was a massive, violent storm. On the afternoon of Wednesday 28 September 2016, severe thunderstorms and tornadoes hit transmission lines in South Australia and triggered a statewide blackout, otherwise known as a 'system black'. System blacks – the complete shutdown of a grid – are extremely rare. The scale and intensity of that storm were always going to cause power outages in South Australia, but losing the entire state was avoidable. At the time of the blackout, the grid was operating as per normal – which means it was supplying the state's demand as cheaply as possible. This is the normal, default setting for grids around the world: no one wants to pay any more than they need to for power.

That afternoon the state's wind farms were generating furiously in the storm, producing 70 per cent of South Australia's local generation, which, at one level, is impressive. It was also standard operating procedure for the grid. Wind and solar were always dispatched into the grid first because they had no operating costs and ran when they could, taking whatever price was set by the market. As the next-cheapest source of supply, the interconnector was also running at capacity. The final top-up of supply (18 per cent) came from two units of local gas generation. The rest of the gas generators were operational, but not switched on. This was the cheapest setup for the grid in South Australia that afternoon. But, as it turns out, it was fragile.

Around 3pm that afternoon, in rapid succession, three transmission lines north of Adelaide got knocked out by the storm. This cut off the west and the north of the state from the rest of the grid. Most of the state's generation, from wind farms, went with it. This separated most of the generators, which were in the north, from the bulk of demand, which was near Adelaide. All that was left were a few wind farms in the south-east, the two small gas units around Adelaide and the power coming in from Victoria. This new 'mini-grid' was suddenly and desperately short of power. It needed extra generation quickly. The remaining wind turbines couldn't provide it as they were already operating at their maximum. The interconnector couldn't dial up any further either. Filling the huge void was simply too much for the two small gas generators operating near Adelaide. The grid shut itself down. The state went dark.

In the post-mortems that followed, much of the debate focused on whether renewables caused the system black. This was the wrong question. The renewables in South Australia did their job that day exactly as they were supposed to, as did the other technologies. This was an inexperience blackout. South Australia went to system black because its grid was not operated to reflect the true risks that day. With the benefit of hindsight, the solution is easy: order the wind farms to reduce output by 40 per cent. Have more gas power stations switched on and operating with plenty of spare capacity, and dial back the interconnector so that both it and the gas turbines are able to respond quickly. But this didn't happen, because the grid was being run on old rules for a new system. Sometimes you have to make mistakes to learn. Renewables can't dial up. They give you everything they have. If you have a high-renewables grid, you need to be able to access more fast-start technology to jump in if something goes wrong. It was a rookie mistake. Without intending to, South Australia had just conducted the world's biggest experimental failure in integrating renewable energy.

Closing power stations

Electricity grids evolve slowly. Most of the equipment is designed to last a long time: poles and wires, meters, transmission towers and power stations. But nothing is immortal. Eventually, old coal-fired power stations, like everything else, become too expensive to maintain and operate and have to be replaced. The basic rule of thumb is that the monolithic coal-fired power stations built since the 1960s can run for roughly 50 years. Sometimes less, sometimes a bit more. The Northern Power Station in South Australia was the ninth coal-fired power station to close since 2012 and the first in Australia to be forced out ahead of its use-by date. Then there was Hazelwood.

Five weeks after the South Australian system black, the French energy company ENGIE announced it was going to close the Hazelwood Power Station six months later, in March 2017. It had been a decade since populist political mechanics started to tinker with Australia's electricity system. Hazelwood was the oldest and second-largest of the four big brown-coal generators in Victoria's Latrobe Valley, producing 20 per cent of Victoria's electricity. Combined, the power stations that had closed since 2012 represented more than 5000 megawatts of generation capacity – about 7 per cent of the capacity of the NEM. The Hazelwood announcement, combined with the South Australian blackout, finally revealed the scale and severity of the mess that had been created. The problem wasn't that an old, inefficient power station like Hazelwood was about to close – it was that there wasn't a plan to replace it.

The effect of Northern and then Hazelwood closing in short succession was material. It hit both the electricity market and the reliability of the system. Without any replacement generation, Victoria had to rapidly increase the amount of electricity it imported from other states. The gas power stations in South Australia and coal-fired power stations in New South Wales cranked up to help fill the void. Even Queensland power stations jumped in, sending more power to New South Wales so it could send more power to Victoria. This interstate cooperation was all made possible by the creation of the national market in the 1990s. But it meant the market was now dangerously tight on supply. The reserve bench of generators was nearly empty.

Wholesale prices increased across the market, reflecting this scarcity. Black-coal generators in New South Wales soon struggled to find enough spare coal to increase their output. Coal production in the state was a tightly managed process, with 80 per cent of production exported through the port at Newcastle. You couldn't just go down to Bunnings and buy a few thousand tonnes. Gradually the generators found the coal needed for extra capacity, but at a higher price. Less coal generation meant gas generators were called on more often, at higher prices than the brown coal they replaced. These new higher wholesale electricity prices were the electricity market calling for new generation to be built. But without the certainty of a proper energy policy in Australia and tight gas markets, the investment freeze remained. Power bills went up, and governments went looking for a way out – or at the very least, someone else to blame.

What's the plan?

Big blackouts can be hazardous for politicians if left unattended. Nine days after the system black, federal Energy Minister Josh Frydenberg convened an extraordinary meeting of fellow energy ministers. Later that day they signed off on the creation of an independent expert review of the electricity sector's reliability. Frydenberg's appointment of Chief Scientist Dr Alan Finkel as the chair proved successful. Finkel was a respected neuroscientist with a doctorate in electrical engineering and experience developing technology start-ups. With remarkable speed, Finkel got his head around the problem and adapted seamlessly when the Hazelwood closure was announced.

Finkel understood that, above all, it was partisan political interference that was the root cause of Australia's electricity woes. In an attempt to defuse this, his detailed report, released in June 2017, anchored his rationale with a foot in both camps: he assumed emissions reductions were necessary and inevitable, but he specifically did not rule out the use of any technology, including new coal generation. He proposed a range of practical measures such as ensuring generators could not close suddenly, the ability to limit renewables in parts of the grid and more transparent grid planning.

Like hiding a child's tablet in a bowl of ice cream, Finkel used a long and sometimes dull list of 50 recommendations to try to camouflage the centrepiece of his proposed reforms: a new Clean Energy Target to provide the confidence needed for investment in new firm capacity. It did not prescribe what should be built or where, only that the net sum of any package of new generation had to meet emissions levels consistent with Australia's international commitments. In theory, you could build a coal-fired power station if you packaged it with enough solar and wind. But just build something. Finkel's review was backed by Frydenberg. Labor was cautious but engaged. A useful coalition of the main business groups, clean and conventional generators, farmers and some environment groups got behind it. Only the ideological extremes rejected Finkel's plan, which simply served as further endorsement.

Federal cabinet found the tablet and spat it out. The CET was rejected by conservatives with a range of fabricated justifications: that it was a 'tax on coal' (even though it had no effect on existing coal-fired generators) or that it was a new, larger renewables target (it actually supported the winding up of existing renewables subsidies). There was no vaguely credible policy case articulated on which to reject the CET – Finkel's attempt to bridge the ideological divide was simply chopped down from one end. It revealed how deep the abyss had become.

Not to be outdone, Frydenberg had been canvassing a plan B proposed by the Chair of the main national energy policy agency. John Pearce from the Australian Energy Markets Commission suggested an even lighter and more palatable solution to find some sort of agreement and get investment moving. A National Energy Guarantee (NEG) would have two obligations: one on reliability and one on emissions. In its simplest form, the NEG provided one mechanism to ensure there were enough renewables built to meet Australia's international commitments, made by the Abbott government in signing the Paris Agreement in 2015. The second mechanism, on reliability, was to ensure there was enough firm generation to go with it.

After nine months of detailed design, discussion, and the same broad and increasingly desperate endorsement from businesses (and helpful opposition from the extremes), Turnbull took the NEG to the coalition party room in August 2018. By then, conservative ideological opposition to anything that remotely resembled a price on emissions had been galvanised as a rallying point against Turnbull. The NEG was no longer about finding agreement on energy policy. It was about taking out Malcolm. Tony Abbott and his cohorts could not be placated no matter what they were offered. Abbott now insisted that the Paris Agreement should not be in the NEG legislation, despite his role in making the original commitment. The emissions reduction target was removed. It was replaced by yet another populist attack on the 'big energy companies' with measures proposed to effectively regulate prices down without addressing the cause of the higher prices: no investment. Turnbull was deposed before the end of the week.

The seventh attempt to reform energy policy in twelve years had collapsed under impossible and impassable ideology. Electricity was no longer a complex machine that served society. It had become a belief. You now believed in coal, or you believed in renewables. Like a bunch of kids tinkering with a car engine, politicians from all sides, all across the country adopted their own pet electricity projects, most without the vaguest idea how the grid worked or what it actually needed. How about a new coal-fired power station in northern Queensland? Or a multi-billion-dollar expansion of the Snowy Mountains hydroelectric scheme? A new solar-thermal power station in South Australia sounds impressive, or maybe a multi-billion-dollar new transmission line? Getting projects up in this new post-technical grid was no longer about efficiency and system requirements, but about how it would look in a media release, and how well you could push your idea in a planning regime overtaken by political lobbying and largesse.

The causes of increased unreliability and higher prices remained unreformed. Having abandoned the policy pathway to fixing electricity, the new Morrison government's answer was to light some torches and conduct a witch hunt of high energy prices. Any notion of the responsibility of governments to lead and solve the challenges faced by society had collapsed. Coal-fired power was no longer a giant machine for making electricity: it had become a symbol of happier times, when electricity was cheap and political correctness didn't ruin everything. Backed by a phalanx of equally ignorant and ideological commentators, the same old group of politicians refused to accept the assertions from electricity businesses and actual experts that new renewables backed by gas were now cheaper than new coal, and easier to build, too. They just stood there in the last days of a dying government, raging against any sort of change with a renewed sense of purpose, however pointless and destructive this was.

Meanwhile, in the background, the grid continued its relentless march towards the end of coal-fired generation. The Liddell Power Station in the Hunter Valley is the next of the 23 remaining coal-fired power stations scheduled to close, with AGL announcing that it will be shut down in 2022. There is still no plan to replace it. If there was any lesson from the last decade, it is that governments don't get to decide if climate change is real or not. They have two choices: design policy that monetises the risk so that businesses can manage the risk and invest, or do nothing. And the latter leaves you with investment gridlock and a country gradually running out of electricity.

4

HOW DOES THE ELECTRICITY GRID WORK?

There's an argument to be made that many of the failures of recent energy policy have stemmed from ignorance – that is, as the debate became increasingly politicised, the people making decisions about what to do with our electricity grid didn't always know, or care, how that grid actually worked. To understand exactly what went wrong in South Australia, and what might go wrong in the rest of the country, we need some understanding of how the electricity market works.

The National Electricity Market (NEM) is a $200-billion machine that produces and sells electricity across five states with relentless precision, 24 hours a day, every day. Like a plane that never lands, the NEM is self-protecting and self-correcting, perpetually trying to navigate whatever gets thrown at it. The Wholesale Electricity Market around Perth is a slightly modified scale model of the NEM. These markets have two jobs: real-time dispatch management and long-term investment planning. In the short term, the market balances supply and demand by coordinating available generators on and off, up and down, adjusting for sudden increases and decreases in demand, units that trip, clouds that suppress solar, wind patterns that die then surge. In the long term, the enduring price trends in the market are designed to create clear incentives to build new power stations or to close redundant ones.

The 24/7 dispatch and management of the grid is largely mechanistic. It will use whatever it has at its disposal at the time to get the job done. Because most of these ebbs and flows are predictable, the market operator can anticipate most situations and find solutions for them. The investment side of the market is more finely balanced and more sensitive to disruption. Investors building electricity generation – be it wind, solar, gas or coal – need to see likely returns on their investment over its working life. The longer-lived the asset, the further into the future investors will look.

Only governments have been willing to overlook these signals and build according to whatever is motivating them, which is normally a highly publicised attempt to fix a short-run political problem: rising electricity prices, reliability, climate change or a combination of the three. But what most governments have discovered, when they've stepped back into the market they used to own, is that it has changed. It's a lot bigger now, and they are no longer the decisive players they were two decades ago. To their chagrin, they discover that they have become no more than attendant lords in the bigger game of the NEM. Most governments never really understood the scale and complexity of the (simpler) machine they were in charge of in the 20th century. It was never their job then, and it sure isn't now.

The energy machine

The NEM is smeared across 5000 kilometres of south-eastern Australia. The same machine that reaches in to run your house also makes aluminium in Bell Bay on Tasmania's north coast, lights the Sydney Opera House and the Adelaide Oval, and powers the trams in Melbourne and the Burnett Heads Lighthouse near Bundaberg. It is a collection of poles and wires, power stations and power points, turbines and transformers variously owned by governments, more than 200 companies and by every home and building owner in the grid. The machine doesn't stop at your fuse box. It reaches into every room and onto every solar-powered roof.

This astonishing piece of macro and micro engineering is a gigantic see-saw. Every second of every day, the machine has to balance the amount of electricity being produced with the amount of electricity being consumed. This requires the precise choreography of an eclectic ensemble of dancers: lumbering coal-fired giants, nimble hydro generators and gas peakers (gas generators that can switch on and off as required), small, short bursts of power from new batteries, the natural surges of wind farms and a growing, ant-like swarm of tiny rooftop solar PV generators. All these have to be continuously matched to a constantly varying demand for electricity: lights switching off and on, coffee machines firing up in the morning, air conditioners surging on very hot days. Big events such as the Melbourne Cup (the race that stops a nation) can occasionally show up as slight dips in demand, but in reality, the supposed national moment of calm is barely detectable. It seems we don't really stop.

The daily cycles of human life define how we use electricity. Like us, an electricity grid sleeps at its lowest demand, which is overnight (streetlights, 24-hour factories, hospitals, the homes of shift workers and children's night lights). It wakes up in the morning to make breakfast, quietens down when everyone goes to work and surges again in the evening around dinner time before gradually settling down for bed. Around 30 per cent of total electricity demand is residential, while the majority is from businesses and factories. But residential demand is much, much peakier than that of businesses. Millions of people use more electricity when they are at home making breakfast and dinner than when they are working in office buildings, schools or factories. Industrial loads are bigger but more stable, while household behaviour drives the spikes and the lulls. Electricity demand is higher during the working week and quieter on weekends. Electricity companies sell a highly weather-dependent commodity. Demand spikes on very hot or very cold days and falls in spring and autumn. There are various tells: the Newport Power Station is an intermediate gas generator just south of the West Gate Bridge in Melbourne. It has an iconic 200-metre-high chimney (topped by *Where's Wally* hoops) clearly visible to passing bridge traffic. When it is blowing steam, Victoria is using a lot of electricity.

This constantly fluctuating balancing act is even more remarkable because electricity must be produced and consumed instantaneously. It cannot be stored. We can store energy in various states, like the potential energy of water in a dam or the charged chemical state of a battery. But once it is converted into electricity, there is no going back. Conducting this national orchestra is the AEMO. It staffs a high-security national control room in Western Sydney complete with walls of big screens and impressive monitors. The AEMO is in charge of planning demand trends, monitoring the weather (for wind and solar generation) and coordinating the activities of all the different generators. But the AEMO is only the conductor. The music everyone plays to stay in perfect harmony is the NEM.

Before the national market, state grids were run like government departments. Each state utility managed its own grid separately, manually scheduling their own stable of local power stations. The bills they sent out were a vague reflection of how much this all cost, depending on the political mood of the day. The creation of a national market in 1998 used the same people and the same equipment but connected it all together and then ran it like a business. (For clarity, the 'National Electricity Market' does not include Western Australia or the Northern Territory. Think of it as you might the National Rugby League or the South Australian National Football League – which, somewhat amusingly, only includes South Australia.)

The market uses price to sort out which power stations get used at any given time. Any generator of any type – from coal-fired power stations, to wind farms to rooftop solar – can bid into the market at any time. Generators are registered with the market operator so that it knows what capacity it has at its disposal. Like a manager of a professional football team, the AEMO will carefully juggle its squad of generators, getting larger firm power stations to conduct upgrades and maintenance during periods of lower demand – typically spring and autumn. If required, the AEMO can even call a power station back into service if it thinks that capacity will be needed.

In this 'wholesale' market, bulk electricity is bought and sold in five-minute blocks, 24 hours a day, in each participating state. The competitive tension in the market is between generators trying to get their electricity dispatched at the best price they can, while the market is trying to find the lowest overall price to supply constantly varying demand. Each active generator bids a range of what they can generate into the market every five minutes. These bids are coordinated by giant computers that calculate how much electricity is needed for every block and dispatches accordingly. The price paid to every generator that gets dispatched is the price bid by the last, highest-priced generator accepted (this isn't as lucrative as it sounds). The staff of the NEM control room monitor this automated process and manually respond (and override) only in extreme circumstances. They also coordinate how and when power is moved between states via the interconnectors to help out with extra demand or get more balance between high and low prices.

In the pre-renewables days, this bidding process followed a relatively predictable pattern. As the grid went to sleep around 9pm, the big baseload coal generators that were on line would dial their turbines down to their lowest setting and bid as low as possible to ensure their power got dispatched. Overnight prices in the wholesale market with large coal generators have traditionally been very low, often below what it costs them to produce the electricity. But getting paid something is better than having to turn off. As morning broke and breakfast demand picked up, the baseload generators would cycle up their output, and prices increased slightly.

Through much of the day, coal would often set the wholesale price of electricity, which meant it never got too high. Prices only tended to jump up if a unit tripped off, or when demand spiked. Gas peakers are multi-million-dollar generators designed to switch off and on as needed. Because they are not working constantly, they require much higher prices to bid into the market. So when demand increased to a level that required extra generation from peakers, the wholesale price spiked too. Once the period of higher demand had passed, the bidding pushed prices back down too low for the peakers to remain on, and coal took over most of the supply. The coal generators that had run at a loss for much of the day made a profit by being paid the higher prices during the periods of peak demand. This is essentially how the 20th-century grid stayed in balance, and stayed cheap.

Because demand changes rapidly and electricity can't be stockpiled, the price in the wholesale electricity market was allowed to be deliberately volatile, skyrocketing to more than $10,000 for a megawatt hour in a summer heat wave, and falling as low as negative $1000 in the middle of a mild autumn night (negative prices mean the retailer or other major customer gets paid to take the power). This volatility was normal; in fact it was a key part of the design of the machine. High wholesale prices were the market signalling for additional generation, prompting sleeping peakers to switch on. Low or negative prices were the market telling everybody who doesn't need to be generating to get out, go home and have a beer. We don't need you now.

These price signals organised the short-term dispatch of power stations. It also prompted long-term decisions about generation: the building of new power stations and the closure of old ones. When prices rose and stayed high in a region over a period of time, that was the market shouting out that it needed someone to build more generation. When prices stayed down, that was the market telling the weakest and oldest of the existing generators to think about retirement.

The volatility, while necessary, posed risks for generators, large industrials and retailers. No one wanted to risk paying $14,000 for a megawatt hour of electricity, nor would a generator want to risk paying up to $1000 to generate one. To hedge this risk, they signed contracts to trade electricity with each other in the future at fixed prices. The 'futures price' for electricity is where 80 per cent of the market gets traded. It's now the main game. This futures market has excluded wind and solar generators because they couldn't guarantee supply in the future. (It's one of the reasons wind and solar developers are so keen to get large batteries installed.) As a result, wind and solar get paid whatever the spot price is at the time they are generating. It can be like riding a roller coaster.

This giant, relentless electricity machine was a success. It found new and more efficient ways of doing the same thing while creating enough stability for companies to finance new generation while redundant capacity was closed. Interstate trading of electricity added competition and discovered an even more efficient way to use the same generators. The NEM became a complex and finely tuned hive mind that worked with remarkable precision and efficiency. It wasn't controlled or directed; it solved. In a recurring theme, you could tell it was a success because, like successful air traffic control or good umpiring, you never noticed it.

When economists were faced with the task of decarbonising this electricity-problem-solving machine, they soon realised the only way was to work with it, not against it. Trying to tell the NEM what to do was like yelling at a swarm of bees. The key to reducing emissions as cheaply as possible was to get the NEM to work it out for you, surprise outcomes and all. Putting a price on emissions was like designing a jigsaw puzzle piece that had to slot neatly into the operation of the market. Pricing carbon didn't interfere with the NEM's delicate balancing mechanisms; it simply made higher-emissions generation relatively more expensive than lower-emissions generation. The detail may have been complex, but the idea was simple: we let the market choose the technologies and where it needs them to be built, but create a situation where it will increasingly prefer lower-emission technology. At a sufficiently high price (somewhere north of $30 a tonne of carbon dioxide), a carbon price would start to push high-emissions (coal) generation out and push lower-emissions generation (gas, wind and solar) in. The idea was to keep going until the job was done.

Unfortunately, by 2009 in the political world, carbon pricing had become a negative, and directly promoting renewable energy was relatively popular with voters. Carbon pricing got the sack. Renewables schemes were hastily repackaged as the A-list policy solution. This was the moment when we stopped harnessing markets to solve complex problems for us, and tried to just order them around instead.

The Renewable Energy Target (RET) required the installation of a specific quantity of a specific technology type. There were no stabilising or firming conditions. In a vague homage to market principles, its primary rule was that the renewables chosen had to be the cheapest. When this blunt tool was programmed into the hive mind, the machine determined that the cheapest technology was wind and the cheapest place was South Australia, which is why it built wind farms there. Lots of them.

Disruption

The disruption of 20th-century markets has been a defining feature of the early 21st century. The disruption is pervasive: landline telephones, video recorders, records and CDs, dating, taxis, media, photography, travel agents, television and retail. In the 20th century, department stores dominated retail. They became iconic brands in themselves, controlling prime real estate and providing unique shopping experiences. Today, many department stores are mortally wounded; their commercial advantages of prime real estate and personal service have become liabilities when competing with online retail. Their business model is increasingly anachronistic: fixed opening hours that lock in labour costs, even as they sell less merchandise. Their disruptors have almost no overheads and can sell online 24 hours a day. Like many other disrupted industries, this is not a neat transition. Consumers still like to visit department stores to inspect the goods and enjoy the shopping experience, they just don't buy so much. It suggests an uneasy and uncertain transition.

It's the same for electricity. Coal-fired power stations are giant electricity factories producing cheap, bulk power. Each power station is a capital-intensive business, generally employing a few hundred people. They help source thousands of tonnes of fuel that needs to be transported, stockpiled and prepared. There are tonnes of waste ash to be removed and dealt with, and large, complex machinery that needs to be maintained. Coal generators are designed to go on and stay on. They can idle up and down within their operating range, but once they are operating, they need to stay on. They are designed to be the dominant player in the market. They are not designed to cope with what happens when they aren't.

By contrast, wind and solar are like online retailers. Large-scale renewables operate in the market in a completely different way to conventional power stations. Wind and solar farms have almost no staff. Their generators have no fuel and are highly automated.

Almost all of the cost of producing wind or solar electricity is in building and installing the turbines or the panels. It costs a renewables generator nothing to switch on. So they run whenever the wind is blowing or the sun is shining. This means they effectively bid in before all other generators, and then take whatever spot price is set. At low levels of renewables generation, their effect was at the margins. They just softened the wholesale price slightly. But as renewable penetration increased, like in South Australia, their impact on the market increased dramatically.

In a small grid like South Australia's, having a growing fleet of wind farms soon caused a major disruption to the state's wholesale market. On windy, mild nights the wind generators started to produce so much electricity that they began to push the existing baseload generators below their minimum levels of output. Some of the surplus got shifted off to Victoria, but the commercial impact of this new, unscheduled generation meant the coal generators started taking a commercial hit. Wind diving in and out of the market kept cutting the coal generators' volumes and shaving their prices. Wind blowing during evening peaks and hot days reduced the number of times the peakers needed to be switched on, which in turn reduced the number of times these firm generators could make a profit.

The Northern Power Station at Port Augusta in South Australia's Iron Triangle was the last remnant of the Playford era of brutalist, state-owned coal generators. To make electricity, it had a workforce of more than 250 people who mined brown coal at Leigh Creek, 280 kilometres away, transported it to the power station and wrangled it through the industrial-scale steam engines. Northern had to cover these fixed operating costs or go out of business. As wind generation entered the South Australian market at scale, Northern's fixed costs overwhelmed what it could earn in this new market environment. And so it closed.

Northern didn't close because it was a high-emissions source of generation; it closed because it was too inflexible. As renewables were forced into the NEM, the market simply solved for this new condition. After the closure of Northern, wholesale electricity prices rebounded sharply as the market switched from oversupply to undersupply. The market was now signalling the need for new firm generation, only more flexible than coal.

The SOS went out, but no help arrived. Cheque books stayed firmly shut. The logical replacement for Northern was new gas generation. Gas generators are flexible and cheap to build but sensitive to fuel prices. The more infrequently a generator operates, the higher it needs the wholesale prices to be. One of the effects of all the new wind farms was a reduction in the frequency of lucrative peak demand events, but an increase in their intensity. In other words, a high-renewables grid changed the market from being constantly slightly volatile, to being occasionally extremely volatile. This marginal business case was further eroded by an increasingly tight gas market in south-eastern Australia: the price of gas had tripled in two years.

Batteries would have been an ideal match for all these renewables, but they were nowhere near the size or price point needed to provide this sort of bulk power replacement.

The machine breaks

One of the enduring political debates in modern Western democracies is around the size and role of government: what services it should provide and what should be left to the private sector. When the NEM was created in 1998, it was founded on the idea that a competitive market would be more efficient than government utilities in providing electricity, and it would enable private investment in new infrastructure, freeing up government funds for other valuable purposes. The role of government had evolved: from service provider to market creator. The market as designed worked seamlessly with the 20th century generators it was designed for.

When Northern closed in South Australia in 2016, it became apparent that the market conditions had materially changed. The traditional investment signals were no longer sufficient. Unsurprisingly, in the wake of the system black in 2016, governments were drawn towards increased intervention. Federal government ministers and the Liberal state opposition started to call for Northern to be brought back to life. Even if it was unclear how they might achieve this, the threat of it added commercial risk to any new firm generation. Not one for half-measures, some federal ministers talked up the idea of the government backing new coal 'somewhere' and keeping Liddell operating past 2022. This kind of trash talk just dialled up the risk.

The RET itself increased risk too. It was harder to predict subsidised intervention, which overrode the conventional investment signals that companies used to forecast market trends. Gas markets had been rendered illiquid and expensive, not helped by governments banning development of new onshore gas fields in Victoria and moratoria elsewhere. In March 2017, South Australia's Weatherill government announced it was re-entering the generation market by buying a diesel-gas generator, and subsidising a battery and a solar thermal power station. The Liberal opposition countered with its plan to fund a new interconnector into New South Wales. The market was now being overridden, first by the RET, then by successive state governments.

Jay Weatherill not only shirtfronted the NEM – he also shirtfronted the federal government. A day after his plan went out, the Premier of South Australia invited himself to the announcement of a virtual power plant project by energy company AGL, staged in the living room of a house in suburban Adelaide. It was to be launched by Federal Energy Minister Josh Frydenberg. In front of the media who had gathered for the announcement, Weatherill simply imposed himself centre stage and then, standing shoulder to shoulder, ripped into Frydenberg live on national television.

It was a memorable moment. Weatherill was facing a state election in a year's time and was about to announce that he wouldn't go down without a fight. To prove it, he started one. Put simply, Weatherill hijacked the AGL event to re-advertise his own $550 million announcement from the day before, and to play the parochial state versus federal card on energy policy. Frydenberg had been critical of the extent of the state's intervention in the electricity market. As Weatherill let fly, a stunned Josh Frydenberg stood uneasily, but politely next to him.

While it all made for great television, the political impact of Weatherill's day out was less memorable. Energy policy and blackouts didn't appear to make any material impact on voters at the South Australian election in early 2018. The primary vote of both major parties was almost unchanged from the election three years earlier. Only a redistribution of boundaries delivered the change of government.

The health of the NEM had more than a feisty Jay Weatherill to contend with. A decade of state and federal government subsidies for rooftop solar PV meant that by the time Northern closed, more than 30 per cent of dwellings in South Australia had solar PV installed. In 2016 it supplied seven per cent of the state's total generation, peaking at 41 per cent of total demand on low-demand, mild, sunny days. Rooftop solar had become a black-ops power station: it couldn't be controlled, but its output kept growing.

The scale of this commercial and investment risk was starkly reflected by the behaviour of AGL, the largest generator in South Australia. AGL had been regularly criticised by governments and regulators for having an overly dominant position in both the retail and generation markets in South Australia. Yet when the opportunity arose, the company did nothing to protect this position. In 2017, AGL announced it was building a 210-megawatt flexible gas peaking generator in South Australia to replace its ageing and creaky 480-megawatt Torrens Island A gas generator. Its larger 800-megawatt Torrens Island B gas power station is expected to turn off around 2026. A thousand megawatts of firm generation had left or was leaving the South Australian market, yet only 210 megawatts were committed to replace it.

A year later, Victorians discovered the same investment paralysis in firm generation after the closure of the 1600-megawatt Hazelwood Power Station. The only thing to happen was a spike in electricity prices and a tightening in demand that ricocheted along the entire east coast. After the general failure to respond to this major withdrawal of capacity, there was no denying that the investment signal in the NEM was broken. The market still dispatched the most efficient mix of available generation, selling contracts and avoiding blackouts if it possibly could. But the breakdown of its investment function was unsustainable. Under the distorted conditions of the post-renewable target NEM, it could no longer initiate new generation when required. It needed urgent repair.

Two separate reviews tried to do just this. The federal government's Finkel Review proposed its considered, almost invasive reform of the market and its arrangements. Its key reform got rejected. The Turnbull government's own Plan B, the National Energy Guarantee (NEG) followed a similar fate. The energy industry watched in disbelief as the federal government kept proposing then rejecting its own reforms, and, as if to emphasise the point, removed its leader for good measure.

Just when you thought things couldn't get any worse, the Morrison government decided to ignore the problems of investment gridlock and climate risk. Instead, it tried to launch a seemingly unlawful and unconstitutional attack on the structure of the Big Energy Companies, playing to dated populist rhetoric, like a rerun of the *Benny Hill Show*. Morrison's attempt to override the courts, seeking to legislate himself powers to order the divestment of energy company assets, was a spectacular way of making a bad situation worse: a new low of government bullying.

The reason this is so serious is that new firm generation is needed in South Australia and Victoria right now, and soon in the other mainland states too. On 25 January 2019, the summer blackout warnings for South Australia, Victoria and New South Wales turned into the real thing for 200,000 Melbourne households, blacked out partly because of a combination of heat and frail old coal generators, but mostly because of political intransigence. Heat waves are predictable. Switching off the air conditioners of 200,000 households in 43-degree heat is a political and technical fail in a modern economy.

If the market isn't fixed quickly, then governments will be forced to build it themselves. The scale of intervention required will siphon billions of dollars from governments and kill the market for good. That decision will mean more than $100 billion that should be spent elsewhere will be diverted into power stations. Australia is currently well advanced on the road to that catastrophic policy fail.

Intervention

The siphoning of government money has already started. Faced with investment paralysis, the fallout of a statewide blackout and a looming state election, the former South Australian Government can justify its intervention in local generation: its $360 million 'government owned' diesel-gas peaker helped avert a blackout in January 2019. Its celebrity Tesla battery (complete with Elon Musk visits) has been profitable and useful in providing power quality services. Its ambitious developmental solar thermal power station may never get built. Still, two out of three isn't bad.

As you might expect an opposition party to do, the Liberal opposition of the time, headed by Steven Marshall, went in the opposite direction, committing $300 million towards a possible new $1.5-billion transmission line between northern South Australia and western New South Wales: a giant extension cord. This new interconnector would be able to move as much as a small power station (around 700 megawatts) in either direction. It would be small in the scale of the New South Wales grid, but big in the scale of the South Australian grid. The claimed benefit was that it would act like another power station, importing electricity from New South Wales into South Australia when needed, and exporting any surplus renewable energy from South Australian wind and solar farms.

Interconnectors are not power stations. They don't generate any electricity, just move it around. Interconnectors are regulated assets: because of their scale and how they're used, there is unlikely to be competition in providing them, so they're treated as monopoly assets – like train tracks and mains water pipes – which means that the process of deciding whether to build or expand interstate transmission lines is made not by businesses but by the chief rule-maker and market defender, the Australian Energy Market Commission (AEMC). Such decisions are complex, even in a healthy market, with many potential consequences. In this case there are plenty of known unknowns for the AEMC to consider.

The interconnector was proposed to be completed by 2021 and would have an operating life of around 50 years. This means it would come into service just as the 2000-megawatt Liddell Power Station in New South Wales was preparing to close in 2022. This closure would tighten supply in New South Wales at exactly the same time when that state was going to be called upon to supply additional generation into South Australia. The Vales Point Power Station in New South Wales turns 50 in 2028, so on current trend it's likely the dispatchable supply from New South Wales is likely to further deteriorate.

The future supply–demand balance in New South Wales depends ultimately on whether there is any meaningful national climate and energy policy in the future. Aside from the possibility of Snowy 2.0 (bulk electricity storage), most new investment in New South Wales is currently in renewables. Without firming capacity, renewables cannot be relied upon to supply South Australia when it needs them. In short, the ability of New South Wales to supply South Australia into the next decade is uncertain.

South Australia faces the exit of 1000 megawatts of firm capacity by 2021, and another 800 megawatts later next decade. The only new local firm supply is Weatherill's 250-megawatt gas-diesel power station and AGL's new 210-megawatt reciprocating engine at Barker Inlet. In addition to the 100-megawatt Tesla battery, renewables developer Infigen energy is building its own 25-megawatt battery in South Australia, plus another 30 megawatt battery run by AGL and owned by ElectraNet. All three batteries only supply small amounts of energy in short bursts. Combined they are not even remotely at the scale needed to firm the South Australian grid in the way an interconnector can.

So what does the AEMC do about the proposed big new extension cord? South Australia is going to need more power over the next decade. Would New South Wales be in a position to supply it when needed? To justify their high capital costs, interconnectors need to be working all the time, not just in peak demand emergencies. If the new interconnector proceeds, it will provide further discouragement to any investors who were thinking of building new firm generators in South Australia. They won't be able to compete with cheaper black-coal electricity from New South Wales or brown-coal electricity from Victoria. If the interconnector gets the green light from the AEMC, then South Australia risks becoming increasingly reliant on interstate generators – like converting your kitchen into a bedroom and getting Uber Eats for dinner every night. With coal-fired power stations already closing interstate, this over-reliance starts to look very risky. But if the interconnector doesn't proceed, South Australia still needs to find at least another 800 megawatts of firm generation from somewhere.

This mess is a perfect distillation of the problem of trying to redesign the grid without a plan. This South Australia–New South Wales interconnector is just one of a number of extension cords proposed by the AEMO. It has developed a general plan for the grid that includes all the possible transmission upgrades, both between states and to new renewable regions. The AEMC has an almost impossible job of deciding whether any of these giant extension cords can proceed without a functioning market and national energy policy. Right now the AEMC cannot predict future investment with any confidence. If it approves the interconnector, it risks chilling the market for new firm generation in South Australia. If it doesn't approve, it risks stranding them.

The repair

The NEM is no longer working as it needs to. The two most recent proposals to fix it – the Finkel Review and the NEG – both land on the same solution: the market itself remains the best way of delivering efficient and reliable electricity, but it needs some repairs. First, there needs to be some formal requirement to reduce greenhouse gas emissions, by whatever level is set by the federal government. Second, the market needs a new mechanism to ensure reliability. The introduction of high levels of intermittent generation has meant this can no longer be assumed. When the market signals it needs more firm generation, there needs to be a mechanism to ensure it gets built.

In 2009, the Rudd government legislated to subsidise higher-cost renewables generation into the grid. A decade later, renewables are cheap enough to no longer require subsidy. We still need to build them, for both economic reasons, because they are the cheapest form of new generation, and to deliver ever-greater emissions reductions. Now the subsidy has flipped. The support is not for the renewables, but for the generation required to back it up. This is a head-snapping reversal of the conditions that confronted the market only a decade ago.

The creation of both an emissions and reliability obligation are the two components of the NEG. This dual certainty is now critical to retaining a decarbonising, stable and reliable grid. Without it, we are sleepwalking towards a much more serious electricity crisis over the next decade. The blackouts have already started.

Australia's chronic inability to sort out a national climate and energy policy is placing increasing stress on the existing governance of the grid as supply conditions tighten, while prices and the risk of blackouts increase. Without a repair of the market, investment in new generation will be partial: mostly renewables, but undersupplied on firming capacity. Governments will feel even more obliged to intervene, cascading the market further and further towards permanent government investment and intervention – re-nationalisation by stealth.

The AEMC and the AEMO have been in increasing disagreement with each other as the conditions within the NEM deteriorate. The AEMO's neck is on the line if there are blackouts and major failures, so it increasingly wants to override the market and back projects that can help out. The AEMC is resisting this. It fears this will cascade into large-scale institutional intervention, creating a regulated electricity market where decisions are made by lobbying and politics, rather than merit and efficiency. The risks here are material. Every time a government or centrally planned asset is built that doesn't deliver enough value to cover its costs, that inefficiency comes back to consumers in higher power bills (or taxes). The AEMO argues that because the market has been so damaged, intervention is required, otherwise the consequences could be dire. They want to buy more insurance, even if it costs more.

As you might expect, businesses that operate in competitive markets tend to back the AEMC, while businesses that prefer a more regulated approach – many renewables developers and regulated poles and wires companies – like the AEMO's approach. These debates are raging right now across the reorganising market. Who pays to connect constrained or remote renewables projects to the grid? Would it be cost effective to pay electricity users to switch off during heat waves? Should we have a market that pays for capacity, rather than paying for the electricity? Who pays for and who builds the new flywheels and other technologies needed to run the decarbonising grid? What are the rules for new in-home batteries? For houses that want to cut the wires and go off-grid? These are just a handful of the technical debates between these expert electricity market bodies – which is exactly where they *should* be thrashed out: by teams of technical and market experts who know what they are talking about. Not by politicians with one-page briefing notes, quoted on the front pages of the national newspapers.

Blackout

To date, the NEM machine has been remarkably effective at avoiding blackouts. In the decade to 2017, the AEMC estimates that less than 0.25 per cent of all blackouts were caused by not having enough generation. The hive mind finds a way. Even now, almost all blackouts are caused by faults in the network: power lines coming down, transformers failing or other equipment problems in the poles and wires. Around three per cent of blackouts are caused by power quality being compromised – where voltage and/or frequency were moved too far from the safe range and couldn't be brought back in time. The system black in South Australia was a blackout caused by an unmanageable collapse in power quality, not by lack of generation.

The common garden-variety network fault blackouts are identified and dealt with by local network crews that move as quickly as possible to physically find and rectify the failed equipment and repair or replace it, like changing a light bulb that has blown. These failures are generally random and hard to prevent, except by spending more money (paid for in power bills) to over-build network equipment or replace it early. This is what is known as 'gold-plating' the networks. Sometimes equipment failure and resulting local blackouts can be triggered by periods of high demand, such as heat waves, when the network is most stressed.

As you might expect, the NEM machine is constantly thinking about how to ensure the lights stay on. It will start setting off alarms as soon as it detects any part of the grid getting close to failing. There are contingencies built into the system, and every generator in the grid can be directed by the market operator to turn on, off, up or down in an emergency. If a blackout does occur, it is almost always controlled. The market operator will direct the local network operator to 'shed load' in order to rebalance the system. This means blacking out areas of the grid until frequency, voltage or demand–supply balance has recovered and power can be restored. It's then up to the local network operator to choose who gets blacked out and who stays on.

Until the closure of the Northern and Hazelwood power stations, the market had plenty of reserves on the bench to manage peak demand events – which occur mainly during summer heat waves. Special preparations are made for heat-wave season: any repairs on generators need to be finished before the start of summer, network operators check their poles and wires, transmission operators double-check their equipment. Heat waves are Grand Final Day for the electricity grid.

The exit of large thermal generation from the market without adequate replacement has left the NEM machine bench-light on, particularly in South Australia and Victoria. This increases the risk of blackouts caused by insufficient generation. This is what caused 200,000 Victorian households to be blacked out in January 2019. The level of risk depends on a cumulative set of events.

First you need a proper heat wave. This is generally when maximum temperatures reach or exceed 38 degrees for three or four consecutive days, accumulating a heat load in buildings. Second, demand is higher if a big heat wave comes midweek: demand tends to dip over weekends, easing pressure on the system, but a heat wave from Monday through to Wednesday or Thursday maximises demand. Stress on the grid is increased if there are high temperatures at the same time in two or more states, particularly South Australia and Victoria: these states' weather systems tend to be strongly related, and the wider the heat wave, the bigger the overall demand on the system.

Demand is higher from late January or February when schools and factories are fully back at work. Light or no winds increases stress, as no wind means no wind generation. Finally, if it's been hot for a while, older generators will have been working at or near their capacity, resulting in increased risk of faults or breakdowns.

These cumulative risks are all factored into the planning for the summer by the AEMO. There is a range of well-established and new measures to mitigate them. Large industrial customers (such as aluminium smelters) are notified in advance that they may be asked to shut down during the peak of a heat wave to free up capacity. Solar PV is almost always working hard during heat waves, and lower voltages across the grid mean it is all getting on: it's a handy black-ops power station, except it still goes home near sunset. Solar PV has the effect of pushing the evening peak later into the hot summer evening, but not actually reducing it by much.

In the last two years the market operator has begun contracting a range of smaller industrial customers, offering to pay them to switch off under high demand conditions (Reliability and Emergency Reserve Trader or the RERT). This isn't as easy as it sounds. Many businesses that sign up to the scheme discover they don't or can't switch off when asked. Heat wave demand is hard to mitigate because consumers value electricity so much more when it's 40 degrees outside.

The most challenging conditions faced by the NEM since the closure of the Northern and Hazelwood power stations occurred in the second half of January 2019. For two weeks, rolling heat across south-eastern Australia kept temperatures high and pushed the capacity of the electricity system to its limit. Potential blackouts from generation shortages were avoided through a combination of enough thermal generation, and wind and solar supplying power at the critical times.

While numbers varied depending on the conditions, around 70 per cent of generation in the NEM for those two weeks came from coal and 13 per cent from gas, while hydro, wind and solar provided the rest. The South Australian Government's backup diesel generators even kicked in, along with small bursts from the batteries at the height of the peaks. Predictably, teamsters for the different technology factions jumped on social media claiming their technology had been the game changer in preventing blackouts.

The eventual blackout of 200,000 Melbourne households on 25 January, a 44-degree day, happened because there wasn't enough generation. Technical problems reduced the capacity of three coal generators in New South Wales and Victoria. Faults are to be expected when running big generators hard in high temperatures. That wasn't the problem. The grid is supposed to have comfortable reserve margins to cover these sorts of contingencies. The AEMO called in every bit of generation it had, pulled in every watt it could find from South Australia, Tasmania and New South Wales. They got major industrial customers to switch off. But in the end it just wasn't enough power to get through the peak. This wasn't a system failure blackout. It was a policy failure blackout.

The biggest risk to reliability will continue to be the things that can't be planned for: units tripping in power stations, bushfires, unexpected network faults. These will continue to be a risk. But the risk of reliability blackouts in summer heat waves is real. The most recent summer of blackouts and blackout warnings is now the new normal. The longer this goes on unresolved, the greater this risk becomes. Until national climate and energy policy is fixed and the NEM machine can work properly again, blackouts in summer heat waves are a constant, escalated risk. The disintegration of electricity market conditions is a slow risk, like climate change. It only gets noticed when power bills spike or chronic blackouts break out.

5

WHAT ARE RENEWABLES?

Australia's electricity system is powered entirely by solar energy. It's just a question of how it gets to us. Fossil fuels are the sun's energy from up to 250 million years ago converted into swamps and forests, and stored as coal, oil and gas. Hydro converts recent solar energy by capturing the rain evaporated from the oceans. Wind turbines convert the movement of air caused by temperature differentials between places that are warmer because of lots of solar energy, and cooler because of less solar energy. And solar panels, self-evidently, convert the sun's light directly into electricity.

Fossil fuels underpinned the industrial revolution and powered the 20th century, and the modern world was built with the cheap, abundant energy they produced. Fossil fuels have four key properties. First, they are energy-dense, which makes it easy to produce cheap electricity. Second, because they are a stored physical resource, the electricity can be controlled and dispatched as required: 'firm' generation. Third, they are finite: fossil fuels will eventually be exhausted or too expensive to extract. Finally, the combustion of these fuels releases a range of pollutants and carbon dioxide – a greenhouse gas – which, unmanaged, will constrain their continued use.

The leading renewable energy technologies to emerge in the 21st century are the opposite of 20th-century fossil-fuelled generators. Wind and solar are energy 'diffuse' and have had to develop major efficiency gains in their design and production to be competitive on cost. They are intermittent, only generating when the wind and sun are available. Renewables are effectively infinite, and they don't produce pollutants or emissions.

The ability of renewable energy to play a major role in the electricity system of the 21st century is only possible because of its accelerated development over the past decade. At the start of the 21st century, renewables operated at the fringes of the electricity system. The industry comprised a large number of technologies under various stages of development, a talking up their promise and potential. The most advanced technologies were expensive and unreliable, while some of the stragglers hadn't managed to produce electricity yet.

Since then two renewable technologies have matured rapidly. The cost of producing electricity from solar PV and wind energy has fallen so quickly that they are now among the cheapest sources of new generation. This has had a devastating effect on most other renewable technologies, which have fallen fatally behind. Many have now all but stopped development. Wind and solar will be the largest types of new electricity investment through to 2030 in Australia. They currently supply around nine per cent of all electricity nationally. This is no longer a boutique industry. These are industrial-scale technologies that have been developing globally for decades and will play a significant role in the electricity grid of the 21st century. So how did this happen? Where did they come from, and why did wind and solar PV succeed?

Solar photovoltaic

Photovoltaic solar panels, known as solar PV, is often cast as a rebel brand. Its brand backstory is mythologised in a manner reminiscent of William Wallace in *Braveheart*: a defiant new technology that is challenging and beating the might of big energy companies, and giving power back to the people. Solar PV power is a game changer, even though it was discovered nearly 200 years ago – by accident – and has been the subject of multi-million-dollar research and development since it was refined in the 1950s – again, by accident. To be a game changer, it needed to achieve industrial scale. To get there, it enlisted the help of major corporations, the military and fossil fuel companies.

Solar PV has been on a remarkable and relentless descent down its cost curve. It has evolved from a lethally expensive, space-age technology into the cheapest source of new electricity on earth. Solar is a true outlier technology. It's the only source of generation that creates electricity without a mechanical process. It converts light directly into electricity without any moving parts. When compared to other generation technologies, solar PV feels like something stolen from the future.

Energy from the sun takes two basic forms: heat and light. The heat from sunlight is used in solar hot-water systems. The energy from photons of light is the other type of solar, and it powers the black, flat rectangular solar PV panels currently on the roofs of more than 2 million Australian households and businesses.

The photovoltaic effect was discovered in Paris in 1839 with two silver-coated platinum electrodes: the effect was observed, but not understood. As a result, the handful of first attempts to replicate the original experiment (which were essentially the first attempts to make solar panels) used similar metals such as gold and platinum. They were not only more expensive, but also produced very little power, so it all seemed a bit pointless given progress occurring in electricity generation elsewhere. After this uninspiring start, there was little interest in solar PV for the best part of a century.

In 1905, Albert Einstein provided the science to explain the photovoltaic effect. Einstein developed a theory about light that became one of the most revolutionary scientific ideas of the 20th century. He identified that light was made up of photons and not waves, as previously thought. When photons hit some materials with enough force, they triggered the emission of electrons. He called this the 'photoelectric effect' (it eventually won him a Nobel Prize). The photovoltaic effect described when these electrons were released into another material, creating a current. Solar PV cells work by capturing this current.

The industrial story of solar PV began in the early 1950s, in the New Jersey offices of Bell Laboratories, the research subsidiary of US telecommunications giant AT&T. At the time, a small team was working on ways of improving the conductivity of the silicon used as transistors in electrical devices (the forebears of the silicon chip). They were experimenting with coating the silicon in different materials when they discovered (by accident) that a specific combination of materials caused the silicon to create a current when exposed to lamplight. Meanwhile, down the corridor, other Bell researchers were looking for better ways to provide the small amounts of electricity needed to supply the remote parts of the growing AT&T telephone network. To this end, they had been experimenting with small wind turbines, steam generators and even the old expensive-metal type of solar PV.

A pair of lead researchers from these teams, Daryl Chapin and Gerald Pearson, were friends. One day in March 1953, they swapped notes. Pearson told Chapin about the silicon configuration he had discovered. Chapin went back and tested it. The coated silicon used as a solar PV cell produced five times more electricity than the 'conventional' metal solar cells he had been working with. The silicon solar cell was born.

By April 1954, the first Bell Solar Battery was presented to the world. The cost of the electricity was thousands of times the cost of mains power: at around US$2000 per watt, a typical 4-kilowatt household rooftop solar system would have cost more than $8 million. The solar battery attracted global media interest but no commercial leads. The only genuinely interested observer was the US Army, which saw a possible role for solar in powering the primitive space satellites secretly under development. In space, the value of being able to generate even small amounts of electricity was literally priceless.

The Space Race unofficially began in 1957 when the Soviet Union launched the Sputnik satellite. In 1958 the US launched Vanguard 1, its second satellite, weighing only 1.5 kilograms and with a central sphere measuring just 16 centimetres across (Soviet leader Nikita Khrushchev nicknamed it the grapefruit satellite). It was fitted with six matchbox-sized solar panels, enabling its radio beacon to send out signals for the next seven years. By comparison, similar battery-powered satellites ran out of power after about two weeks.

Vanguard 1 (which is still orbiting the earth right now) was a stunning technical and PR success for the US space program and solar PV technology. By 1972, around a thousand Soviet and US satellites ran on solar power. Nearly two decades of space racing had increased panel efficiency and reduced cost. By the start of the 1970s, the cost of solar PV had fallen from US$2000 to around US$100 a watt. Though much cheaper than in 1954, at these prices, a 4-kilowatt household solar system would still cost more than $400,000: not competitive for terrestrial applications.

In 1969, the global oil company Exxon was reviewing its long-term strategic future: where would it be in the year 2000? Oil reserves were finite, and Exxon decided it should start looking further afield. One option was to explore new energy technologies such as solar PV. Exxon executives believed the price of electricity supplied had to fall significantly for solar PV to be commercially viable on Earth. They created the Solar Power Corporation (SPC) to drive down the cost of solar PV. Space-Race solar PV cells were being produced using the 'money is no object' approach. SPC introduced some common sense to the manufacturing process for solar panels, sourcing scrap silicon from the semiconductor industry and skipping some expensive production processes. This brought down the cost from US$100 a watt to around US$10 a watt. And at that price, solar was a commercially viable proposition.

Solar began to find niches in remote, off-grid locations. It was ideal for powering the lighting on Exxon's buoys and oil platforms, leading other oil companies to start their own solar PV development businesses. By this point, most solar cells were being wired together and fixed onto frames for convenience of installation and protection.

In telecommunications, microwave repeater stations had started to transmit phone signals over long distances, only needing small amounts of power. By using solar PV in remote locations, telecoms could move beyond the range of electricity grids, opening up telephone services to far-flung communities. Gough Whitlam created Telecom Australia in 1975 and required it to provide telephone and television services throughout Australia 'as far as was reasonably practicable'. This meant getting telephones and TVs into remote towns hundreds or even thousands of kilometres from the electricity grid. Ipso facto, Telecom became one of the world's biggest customers for solar-powered remote telephone systems.

By 1978, the first chain of 13 solar-powered microwave repeaters connected Tennant Creek with Alice Springs, 500 kilometres away. Through the 1980s, Telecom continued to roll out another 70 solar-powered remote systems across the continent, becoming a global testing ground for the durability and reliability of commercial solar. This Australian experiment was important: it showed that solar PV could deliver reliably and cost effectively in harsh conditions. This encouraged further research and efficiency improvements, opening up new markets. By the 1990s, dozens of solar PV research facilities had been established around the world, in universities (including the University of New South Wales and the Australian National University) and by a range of electronics and energy companies. By the late 20th century, there were different types of solar PV technologies competing on two key metrics: reducing the panel cost and increasing the amount of electricity it produced (known as the conversion efficiency – the amount of available sunlight converted into electricity). These different tribes of solar fell into three broad groups.

Crystalline: These are the black solar panels bought by most Australian houses, and they make up 94 per cent of the global solar market. Made from crystalline forms of abundant silicon, they have continued to fall steadily in cost while improving in performance. Crystalline silicon is now the cheapest source of new utility electricity.

Multi-junction: The supercars of solar. Made using compounds such as gallium arsenide, they can reach conversion efficiencies of up to 46 per cent, but at a much higher cost. Most solar panels can only convert photons from part of the light spectrum, which reduces their conversion efficiency. Multi-junction panels use layers of different compounds to capture more of the light spectrum.

Thin-film: Thin-film solar was installed into watches and calculators in the 1970s and '80s. It's cheaper to produce than traditional cells, but has historically had a lower output. Though the performance of thin-film has approached and even surpassed some crystalline solar, this is offset by a shorter operating life, as the thin coatings tend to degrade faster. Next-generation thin-film includes dye-sensitive perovskite and polymer cells with potentially limitless applications coating building products. But this isn't as easy as it sounds – if a roof tile lasts 30 years but the solar coating lasts only five, what do you do?

Solar PV was commercialised as a remote power source, not a solution to climate change. But it was adaptable. By the start of the 21st century, US, Japanese and European solar-panel manufacturers started to win contracts to build large-scale solar farms, underwritten by renewables subsidies. The zero-emissions electricity generation business soon dominated sales. But the power it produced was still expensive compared to grid-supplied electricity. In the 2000s, the emerging solar PV industry was catapulted towards grid-parity pricing by a new entrant that would change the game permanently: China.

In 2005, China was industrialising so fast it was struggling to supply its economy with enough electricity. At its peak, China was already building a coal-fired power station a week, and also increasing gas and nuclear generation. Hungry for new capacity, the Chinese government passed a renewable energy law and launched a massive escalation of its manufacturing of renewable technologies, supported by generous financing and tax incentives. By 2010, China had taken over the global solar PV market, producing around 50 per cent of global supply. Chinese manufacturers built huge solar panel factories, backed by guaranteed Chinese contracts. The surplus capacity was then earmarked to compete in the growing subsidised solar markets in Europe and the US. As with most manufactured goods, Chinese panels were significantly cheaper than their western rivals. Once mobilised, cheap Chinese solar drove most of the established manufacturers to the wall.

At the same time, the silicon used to make solar got cheaper. Until around 2005, the solar PV industry had continued to source most of its silicon from semiconductor scrap. But continued growth in the size of the solar PV market meant there was no longer enough scrap to meet demand. Scarcity of silicon supply for solar sent prices skyrocketing, which in turn drove investment in new dedicated specifications of crystalline silicon that didn't compromise output. By the end of the decade, the new silicons were cheaper than the semiconductor scrap they had replaced. The relative abundance of silicon was critical. It enabled the rapid scale-up of solar production without creating scarcity price increases in its supply chain.

Solar PV costs have continued to fall. Solar PV was US$2000 per watt in 1955. Utility-scale solar is now below US$1 per watt, which puts its electricity costs below new coal or new gas. This is without any price on greenhouse emissions. The electricity from residential solar is nearly three times more expensive. Solar PV is now a multi-billion-dollar industry, with new variants of solar capable of reducing costs even further. Its global scale-up is astonishing: 2 gigawatts installed globally in 2002, 100 gigawatts in 2012, 400 gigawatts in 2017. Most new large-scale electricity generation projects slated for development in Australia are now solar PV.

As solar continues to increase market share, it will require a similar scale-up of supporting technologies and generation to smooth out its erratic tendencies. It is only a partial replacement for existing coal-fired generation. But solar is not a fad. It is a game-changing way of making electricity. And it will be at the centre of the 21st-century electricity grid.

Wind

You could say that wind and solar PV are the odd couple of renewable energy. Solar PV is about the invisible actions and reactions of sub-atomic particles, the industrial wing of the computer age. Wind energy is old-school. It's physical generation using gales and giant turbines. For thousands of years, wind has been harnessed to grind grain and sail the oceans. Once electricity generators were invented in the 1880s, they were soon hooked up to the back of a windmill. These first wind turbines produced electricity, but it was more expensive than the steam engines developed for the job. By the start of the 20th century, wind generators found occasional uses in off-grid applications, predating solar by more than half a century. It was a fringe generation technology.

Wind energy's breakout moment came as a result of the two oil-price shocks in the 1970s. When the Middle-Eastern oil producers turned the taps off, the developed world started to look more aggressively at diversifying its energy supply. The US national government, and some US state governments, including California, passed laws from 1978 encouraging increased use of 'alternative' energy sources, including wind energy. At the time, there wasn't a wind industry to speak of, though the Danes had been tinkering with and installing small turbines since the start of the 20th century. When American customers arrived in Europe looking to find someone to build them some wind generators, a small group of Danish agricultural machinery companies put their hand up. Two of these businesses, Vestas and Danregn (now Siemens), evolved into two of the biggest wind turbine companies in the world.

The refinement of utility-scale wind turbines was an exercise in adaptation rather than invention. A wind turbine is the combination of two established technologies: windmills and generators. The generator is located directly behind the blades, connected via a gearing system in a housing called the nacelle. Refinement involved working out the most efficient number of blades (three), how to reduce their weight using fibreglass and alloys, and how to make the turbine bigger to improve efficiency. The turbines also needed to be able to reduce power in very high winds, or risk being destroyed. The solution was blades that automatically feathered into the wind at speeds above 25 metres per second (90 kilometres per hour).

Wind turbines in the 1980s were small by modern standards, rated at around 60 kilowatts each with a blade diameter of around 25 metres. Typical wind turbines installed in Australia today have around 40 times the output. Some larger offshore wind turbines reach up to 8 megawatts each (more than 100 times more power), with a rotor diameter of around 160 metres. The small early wind generators were more expensive than grid electricity, so most were installed with the support of renewable energy schemes: first the renewables subsidies in California in the early 1980s and then in Europe. Denmark and Germany took particular interest in using renewables policy to support development of local wind turbine manufacturing industries. Denmark, regarded as the wind energy capital of the world, sources around 44 per cent of its electricity from wind, relying on nuclear power from Sweden and coal and gas from Germany when it's still.

Claims about renewables creating jobs are strange, because a wind farm employs almost no staff, in the same way that it has no fuel costs. Apart from occasional maintenance and repairs, there isn't much for anyone to do – and besides, keeping staff numbers down keeps costs down. The location and size for each wind farm are based on four key factors.

1. Commercial: In most cases it is cheaper to produce the same amount of electricity with fewer, larger turbines. Regions with relatively higher electricity prices (such as South Australia) attract more projects because they earn more money for the electricity they produce.

2. Grid: Wind farms try to locate on or near the existing grid, preferably where their electricity won't be competing for access with other nearby wind projects (which will be generating at the same time).

3. Wind: Wind farms work best where there are strong, constant winds. The roaring forties blow unimpeded across the Southern Ocean and are a world-class wind resource. Further south tends to be better for wind in Australia (further north tends to be better for solar). To maximise this resource, wind farms are often located near the top of ridge lines, which run perpendicular to the prevailing wind. The land form acts as a funnel, compressing and accelerating the wind, which improves performance. Wind farm sites can't be too rugged or inaccessible, as giant specialised cranes need to be able to get to the sites for installation.

4. Community: Finally, and importantly, wind farms need the support of the local community. A wind farm is a series of large, rotating machines up to 200 metres high, located in regional

Australia. The landholders hosting the turbines typically receive an annual fee for each turbine on their land, while their neighbours have all the proximity with none of the benefits. The quality of life some attribute to living in these regions can feel compromised if large wind turbines start spinning nearby. In some extreme cases, the resulting annoyance and anxiety have led to claims by wind turbine neighbours that they suffered physical illness as a result of infrasound or low-frequency sound waves reported to be emanating from the turbines. In the wave of pseudoscience that has followed, some advocates for sufferers of what's called 'wind turbine syndrome' called for wind turbines to be built no closer than 10 kilometres from dwellings. That's the distance from Manly to the Sydney Opera House, or from the Melbourne Zoo to St Kilda beach. It would protect a superhuman level of hearing.

Some of the best sites for proposed wind farms happen to be in the vicinity of rural properties owned by powerful and influential Australians such as entrepreneur Dick Smith, broadcaster Alan Jones, media magnate Rupert Murdoch, former ABC Chairman Maurice Newman and the family of AFL CEO Gillon McLachlan. Some, like McLachlan, confined their opposition to legal action in an attempt to prevent a South Australian wind project proceeding (without success). Others have been vocal critics of renewable energy. This influence shouldn't be confused with conspiracy. Most wind farm projects proceed through the normal planning and approval processes. Some are delayed; others are modified. They do or don't proceed according to their actual planning merits. For all the conspiracy theories over the years about secret plans and multi-million dollar funds to undermine and stop renewables projects proceeding, there is no evidence that this has occurred, nor has it been successful.

Another concern raised by anti-wind-farm campaigners was their impact on bird populations. In 2006, the Howard government infamously blocked a wind farm development in Victoria because of its threat to four species of migratory birds, in particular the endangered orange-bellied parrot. Research into human activity causing bird deaths later found that cats, power lines, buildings and cars posed a far greater threat to all bird species than wind turbines. The wind farm in question eventually got completed in 2015.

Wind turbines can be built offshore. However, as the turbines still need to be anchored in the seabed, these projects are more feasible where the water is only a few metres deep. The ocean floor around Australia's coastline tends to fall away quickly. Every extra metre of water depth increases installation costs. It is also theoretically possible to build floating wind turbines, but they are expensive and are at risk of damage from storms. A small trial floating wind farm has been operating off the Scottish coast since 2018. For both floating and fixed turbines, there are additional costs of maintaining the turbines at sea. As a result, offshore wind farms in Europe tend to use larger and more durable turbines to reduce maintenance costs.

The first wind turbines in Australia were located in smaller remote grids, where the higher cost of wind turbine power was more cost effective in displacing diesel-fuelled generators. Synergy in Western Australia built Australia's first experimental wind farm at Esperance in 1987. Then ETSA built the second at Coober Pedy in the far north of South Australia in 1991, then Hydro Tasmania on King Island in Bass Strait in 1998. When the Mandatory Renewable Energy Target (MRET) began in 2001, wind produced the cheapest electricity of the new technologies. It has remained the largest source of renewables generation outside of hydro, and current modelling predicts it will continue to lead new generation investment through to 2030.

Like solar, wind energy is intermittent. The wind broadly follows seasonal and daily patterns, but is also stochastic within these trends. This randomness differentiates it from solar PV, which has (obviously) a highly predictable daily generation pattern interrupted by cloud cover. Electricity market operators include detailed weather forecasting as part of their day-to-day operations and can accurately predict renewables generation output. As wind and solar's share of generation increases, the grid will become even more sensitive to changes in solar irradiation and wind speeds. In other words, the Bureau of Meteorology will play an increasing role in the operation of the grid.

The cost of wind energy has been steadily falling over the past two decades. The cost varies depending on each wind farm location, the turbines used and the contracting parties. But wind energy is commonly contracted at around $60 per megawatt hour, which is on a par with large-scale solar and below the wholesale cost of electricity in the NEM. The reason for this is a combination of economies of scale, technology improvements (in particular increasing turbine size) and increased competition between wind turbine manufacturers in Europe and Asia.

The randomness of wind energy does not mean it can be made ubiquitous. One of the early, vague arguments used by proponents of wind was that if you built enough wind farms far enough apart, then the wind would always be blowing somewhere, sometime. In other words, the theory went, enough installed wind could act as a firm generator. The market operator took the idea of firm wind so seriously it has even ascribed a value to it – estimating that around nine per cent of installed capacity would always be spinning in summer, seven per cent in winter. This firmness theory was dented on 7 July 2016 when the 1200 megawatts of wind turbines spread across South Australia produced only 20 megawatts of electricity (less than two per cent of capacity).

As renewables generation increases across south-eastern Australia, there will be times when strong wind or solar in one part of the grid helps cover for still and dark conditions elsewhere. But this cannot be guaranteed. Wind and solar are partially complementary, in that they are powered by different systems. Sometimes the wind will blow at night. Sometimes the sun will shine brightly when it's still. The variability also means there will be times, particularly in winter, when there is no wind and little clear sunlight for days. The Germans have invented a word for this: *dunkelflaute* (dark and still). Managing *dunkelflaute* will be one of the big challenges of the 21st-century grid.

Hydroelectricity

Long before wind and solar started to push their way into the grid, hydroelectricity was providing affordable, zero-emissions, utility-scale electricity around the world. It remains the golden ticket of electricity generation and is still the leader in renewable energy, supplying around 16 per cent of the world's power. Hydro is cheap, it can be turned on and off at will, it's replenished every time it rains and it doesn't emit any pollutants or gases. When it rains in Tasmania and the Snowy Mountains, it rains electricity. Hydro's share of Australia's total electricity output varies between five and nine per cent, basically depending on how much it rains.

The energy in rivers had already been used to turn millstones for centuries before the discovery of electricity. Like windmills, the first electricity generators were soon hooked up to water mills. Water's greater energy-density made it easier to convert into electrons.

Hydro became one of the first types of generation developed when electricity grids emerged at the end of the 19th century. The first hydro generator in Australia came on line in 1891. Westinghouse Electric built the first utility-scale hydro power station in 1895. Three months later, the Launceston Council commissioned its first hydro power station at Duck Reach. In 1916, a new hydro generator using the water in Tasmania's Great Lake began supplying electricity to Hobart, pushing out the small coal-fired generator in town. From then on, hydro became the power source that underpinned the expansion of the Tasmanian grid through the 20th century.

Conventional hydroelectricity uses gravity to convert the potential energy of dammed water into electricity. The water is funnelled downwards at the base of the dam wall through a penstock to spin turbines. The amount of power produced is determined by the distance of the drop and the volume of the water. The further the water drops, the faster each turbine is driven and therefore the more energy they produce. As a result, most of the world's leading users of hydro (Austria, Norway, Sweden, New Zealand, Canada, Brazil) have lots of mountains and lots of rain (or snow). The low cost of electricity has put hydro dams high on the list of economic development projects in poorer countries with the right geography.

Hydroelectricity can have a downside. Building dams on rivers has major local and downstream environmental impacts. The world's biggest hydro generation project is the Three Gorges Dam completed in 2003 in central China. It dams the Yangtze River upstream, creating a million-square-kilometre lake to produce 22 gigawatts of electricity capacity (nearly half of the total capacity of Australia's National Electricity Market). At the time it was being constructed, the dam was opposed by many environmentalists because of the destruction of thousands of square kilometres of habitat for fish, plants, birds and mammals. The Ethiopian government is halfway through completing its Grand Ethiopian Renaissance Dam across the Blue Nile river. When completed, it will take a decade to fill. The Egyptian government is opposed to its construction, fearing the dam will reduce the amount of water flowing into the Nile and therefore into Egypt.

Hydro is a fast-response generation technology, making it highly suited to backing up intermittent technologies such as wind and solar. The lakes and dams of the Snowy Mountains and across central Tasmania are basically giant batteries. They are the equivalent of millions of kilowatt hours of electricity, waiting to be dispatched, rapidly if required. Hydro capacity is constrained by simple geography. Australia gets around seven per cent of its electricity generation from Tasmania and the Snowy Mountains hydroelectric scheme, plus a handful of smaller dams along the east coast. There are effectively no more sites where hydro generation can be built (except maybe the Gordon-below-Franklin Dam in Tasmania, but let's not go there again).

Wave and tidal

Since the 19th century, inventors have been fascinated by the motion and power in waves and tides. Two thirds of the Earth is covered with this constant, clean source of kinetic energy. Waves are caused by the effect of wind on the water's surface. Tides are caused by the gravitational pull of the sun and moon. Wave and tidal power technologies seek to convert this energy into electricity. After more than a century of research, the noticeable absence of these types of power stations suggests this is a trickier proposition than it might at first seem.

Most of the main tidal power technologies are derived from hydroelectricity. The theory is simple enough: the tide causes a body of water to rise, the water is trapped behind a barrage, and then its energy is extracted as it is released. Perhaps the most iconic of these possible projects is the Severn Barrage in Wales. The mouth of the Severn River has a tidal range of 14 metres, enticing multiple attempts to build a generating barrage over the last century. The idea even received multi-million-pound, UK Government–backed feasibility studies in the early 21st century. It appears likely to remain a concept. Even with the second-biggest tidal range on earth (the biggest is in the Bay of Fundy in Canada), the amount of energy produced cannot justify the project cost.

The idea of electricity from wave motion attracted similar interest, and was one of the new technologies developed following the oil crisis in the 1970s. Because the wind creates waves, most of the energy is at the surface – anyone who has swum in the surf knows it's much easier to swim under a wave than through its crest. The technical debate in the wave energy world has been pretty simple: the closer to the surface a technology can operate, the more energy it can harness, but the greater its risk of being damaged. The cost of wave energy technology is exacerbated by the corrosive and invasive nature of any attempt to generate electricity in the ocean environment. It's a tough place to do business.

There have been more than 100 wave-energy technologies deployed around the world, many looking like they came from the set of a *Star Wars* film: giant steel snakes bobbing in the ocean, floating batwings and swaying submerged steel lollipops. Australia had two main wave technology companies, Oceanlinx in Port Kembla and Perth-based Carnegie Wave Energy. Oceanlinx developed an oscillating water column that sat on the surface of the water and used the wave energy to drive a piston, which then turned a turbine. After nearly 20 years of researching the technology, Oceanlinx went into receivership in 2014, leaving one of its devices stranded in the water off the coast of South Australia. Carnegie has been developing its technology of submerged buoys tethered to the ocean floor for more than a decade. As the buoys sway with the waves, they pump sea water to the surface, and this kinetic energy is then used to turn a turbine. In 2014 Carnegie won a contract to build a demonstration plant for a naval base and continues to develop the technology, although it has now diversified into solar PV and batteries.

Most tidal and wave energy companies have closed or diversified over the past decade. They have not been able to progress anywhere near the speed of wind and solar, and in the case of wind, wave energy looks increasingly like a more expensive way of harnessing the same resource. The wide range of technologies being researched suggests wave and tidal energy technologies were only at the start of the development pathway, stuck in first gear as solar PV and wind raced further down the cost curve. It's a long way back.

Bioenergy

The term 'bioenergy' describes solar energy trapped in plants (fossil fuels minus 250 million years), the electricity derived from the combustion of organic matter. Like hydro, its use in electricity generation is mature and constrained by the availability of suitable fuels. In Australia, these come from three main sources: the waste cane from sugar milling, methane gases that accumulate in landfills and the woody wastes in paper manufacturing. Bioenergy is a proven electricity technology and produces firm generation, making it more valuable in the future. But its finite fuel sources constrain its ability to scale up.

Sugar mills in Queensland and northern New South Wales originally used the waste cane (bagasse) to heat their boilers. By the 1950s, the technology emerged to expand this to power small electricity generators, first for use by the mills and towns, then sold back into the grid. Electricity from bagasse contributes around one per cent of electricity supply in Australia.

Landfill gas generation evolved towards the end of the 20th century as a by-product of stricter guidelines about the design of landfills and tighter regulations preventing the uncontrolled release of landfill gas (methane) into the atmosphere. Modern landfills turned decomposing household garden and food waste into bioreactors. Smart businesses such as Energy Developments Limited and LMS soon developed containerised mobile power stations that could be parked on a landfill, extract the landfill gas and convert it to electricity, sell it to the grid and move on to the next landfill when depleted. Bioenergy will continue to make its small but stable contribution to the electricity system. In Australia, the idea of expanding bioenergy generation by growing crops for energy is constrained by cost and competition for arable land and water. There are better things to do with these finite resources than grow fuel.

Geothermal

This energy sources heat from below the Earth's surface, the remnant energy from when the Earth was formed and from gradual decay of radioactive materials kilometres below the surface. This heat occasionally appeared naturally on the Earth's surface as hot springs and geysers. As electricity grids developed around the world, techniques were developed to convert this heat for electricity generation. More than 20 countries including the US, New Zealand

and Iceland generate electricity from their geothermal resources. Globally, these sites have typically been found near the edges of tectonic plates, where the heat gets closer (around three kilometres) to the surface, tapping into hot aquifers that carry the heat to the surface (via water). Geothermal power has almost no greenhouse gas emissions, and it can provide constant baseload generation. However, while it is zero-emissions generation, geothermal may not be so 'renewable' after all: some larger and more mature fields in the US are now discovering the heat source is gradually being depleted after decades of energy extraction.

Australia's infamous expedition into geothermal energy occurred in the 2000s. Geologists identified a number of large resources of hot, dry granite around five kilometres underground in and around the Cooper Basin in north-eastern South Australia. The big difference between these resources and more conventional geothermal power was the absence of water, which made energy extraction harder, as in conventional geothermal power water carried the heat to the surface.

Because the Australian heat resource was dry, the idea was to drill two deep wells into the rock and pump water down the first where it would become super-heated and pressurised then return to the surface in the second. This pressurised steam could then be used to generate electricity. Put like that, it sounds so simple. In 2007, the developers of this 'hot fractured rocks' technology were so confident the technology would work that the industry frontrunner, Geodynamics, appointed a power-station engineer named Gerry Grove-White as its CEO. They were already planning for the power generation, assuming the heat extraction was in the bag. The company had the backing of Origin Energy and big plans for clean generation.

As climate policy debate accelerated from 2006, economic modellers started to include geothermal energy in their forecasts, even though the technology hadn't been proven and no power had been generated. All the major electricity retailers invested millions in separate geothermal businesses. The major geothermal companies listed on the Australian Stock Exchange to raise capital for the expensive drilling and exploration. Their share prices started to skyrocket; geothermal was the 'next big thing'. In 2008, it was forecast to supply around a third of the RET. In 2011, the Gillard government's Treasury modelling assumed geothermal energy would supply up to 23 per cent of national electricity by 2050! Thousands of mum-and-dad investors bought a share of this clean-energy success story. More than $800 million was raised and spent.

Eventually, the hype around the hot rocks cooled. Geothermal had massively over-promised and under-delivered. Australia's hot rocks were deeper and in different rock formations to those that had been exploited successfully elsewhere. At those depths, the conditions were intense. Drill rigs and bits were wrecked. The cost of such deep holes was expensive – more than $10 million each. Eventually, Geodynamics managed to get a small trial generator running, but the cost of each electron generated was astronomical. A cold calculation of the resource concluded it was not economically feasible, and it was closed in 2016. The heat is still down there if better technologies are one day found to extract it. But right now, geothermal is not part of the future electricity supply in Australia.

Lessons

Some people probably still see renewable energy technologies as gimmicks: the industrial equivalent of a dream catcher or fixie bicycle. This thinking is way out of date. Renewables have become part of the industrial fabric of the global economy, built and installed by corporations, funded by global capital and used to help power major economies. After decades of investment and subsidy, wind and solar PV have broken the research-and-development shackles and their cost is now falling, backed by huge economies of scale.

The most recent reviews of different electricity generation costs from the CSIRO, the Finkel Review and the US Energy Information Administration all find that megawatt hours produced by wind and solar generation are now the cheapest or among the cheapest of any new electricity technology. These costs are without any subsidy, and without a price or constraint on emissions, but they don't include firming costs. No matter how old you are, wind and solar will play a major part in the supply of electricity for the rest of your life. The success of these technologies has come at the expense of other less-proven renewables, which have struggled to retain funding as they fall further behind on the cost curve.

The emergence of solar PV after more than half a century in the lab is a useful reminder that technologies can take time to break through and over-deliver. The failure of geothermal in Australia is a reminder that the opposite is also true. The industry oversold its potential, and was not subjected to sufficient scrutiny from governments or investors. Its reliance on equity markets for what was, in reality, high-risk research and development exposed the weakness of Australia's venture capital funding and made life much harder for anyone wanting to follow the same path. As we look to emerging technologies to help supply firming generation for wind and solar, it would be prudent to not make the same mistake twice.

6

WHAT IS FIRM POWER?

After the system black in South Australia and the closure of the Northern and Hazelwood power stations, the political narrative shifted abruptly. Gone were the heady days of more renewables and falling electricity prices. The political post-mortem was abrupt. Suddenly it was all about securing the system: everyone started talking about baseload power, how to replace it and how to get power bills back down. Politicians rushed to adopt their own pet projects based on similar criteria to those that had created the mess in the first place: does this look good in a media release, is there a photo opportunity, can I say the word 'baseload' a lot?

As a result, by the end of the year, Australians were trying to comprehend a whole suite of new technologies and solutions being pitched to them. Government tenders were run in secret; financial estimations were redacted; invisible investors were apparently queued up, shovel ready to build; actual costs and deals were hidden by quoting meaningless price commitments; and due diligence became a kangaroo court. The bottom line was that Australians were getting a whole bunch of politically opportunistic electricity projects, whether they were a good idea or not.

All this commotion had no effect on wind and solar PV projects, which continued to pedal away backed by the Renewable Energy Target (RET). Renewables investment had been briefly becalmed by the Abbott government's review of the RET in 2015. Despite Abbott's previous and subsequent rhetoric on climate change and coal-fired power, his delivery as prime minister of a common-sense outcome on renewables brought important confidence in the future of the scheme. It was simply dialled down to reflect the drop in demand since the original 45,000-gigawatts-per-hour target that was legislated in 2009. As a result, here is a sentence you never thought you would read: the Abbott government contributed to a renaissance for clean energy in 2017 (along with sustained falls in the cost of wind and solar PV and higher wholesale prices post-Northern and Hazelwood). Renewable projects were flying into the grid at such a pace that by 2017 the regulator predicted the target would be oversubscribed.

In the same year Chief Scientist Dr Alan Finkel delivered his advice to the Turnbull government, confirming the seismic shift in electricity economics: wind and solar were now so cheap they were setting the price for new electricity generation in the grid. Until that moment, renewables had been seen as a more expensive generation technology imposed upon the grid by government fiat. Now they *were* the grid. Investment thinking shifted 180 degrees: the starting point for new builds was renewables generation, and the next question was how to firm it so generators could sign contracts for guaranteed supply in the future. The core design elements of both Finkel's Clean Energy Target and the subsequent National Energy

Guarantee (NEG) picked the same mark: how do we subsidise firm generation to support renewables?

Firm or dispatchable power is a generator that can be turned on and off as required. It can be anything – hydroelectricity, coal, gas, a battery, pumped hydro, bioenergy – that has an on and off dial and can be adjusted up and down when the wind dips or the sun stops shining. As the scale of intermittent generation increases, flexible firm generation is required to manage their intermittency. Less-flexible 'baseload' generators – such as coal and nuclear – cannot adjust from off to flat-out, to off again. The more renewables are used, the more flexible the firm generation needs to be.

Baseload power

Since the first grids in the late-19th century, electricity supply has been anchored to large, controllable engines, whether powered by coal, gas, hydro or nuclear. This led to the term 'baseload power', which has been subjected to chronic misuse ever since. Baseload power does not mean 'the minimum power requirement of the grid'. Nor does it mean electricity that is firm or dispatchable. It is simply a term used to describe a certain style of 'go on, stay on' power generator: principally coal or nuclear. Coal-fired power stations can take days to start up. Once running, they are best run continually for long periods of time, where they will deliver a constant output within a prescribed range.

This 'baseload' supply is ideally suited to meet the demand of large industrial customers such as aluminium smelters. It's not especially suited to residential or commercial demand, which varies significantly from day to night, weekday to weekend, summer to spring. That's why baseload generators such as coal and nuclear require the help of more flexible generation to manage short-term spikes in demand. One of the reasons we have overnight off-peak water tariffs is to try to use up all the surplus electrons being produced by coal-fired generators at 3am, after they have dialled down to their lowest output.

In technical terms, the grid is indifferent to the source or type of electrons produced, providing there is enough generation to meet demand and there is sufficient regulation of power quality to regulate voltage and frequency. To deliver this constant bulk power supply with increasingly intermittent wind and solar will require new, highly responsive generation (or large-scale storage). No prescribed minimum level of baseload generation is required. Flexible, firming generators will need to be introduced gradually, walking the street between new renewables and old coal. The ageing coal generators will exit one by one over the next 20 years, like the von Trapp children singing goodbye. As a result, Australian electricity grids will still have 'baseload' generation in some form for decades.

Which takes us back to Turnbull and Finkel. Radical ideas are challenging for most people, even if delivered by the Chief Scientist and backed by a broad coalition of industries. In 2017 the public debate fractured. Incredulous conservatives dismissed the thesis out of hand and doubled down on demanding coal-for-coal replacements. Malcolm Turnbull could feel the ground parting beneath his feet, and his opponents rallying. He amped up the electricity pantomime: introduced outraged attacks on electricity retailers, bullying AGL to defer its decision to close the Liddell coal-fired power station in 2022, the next von Trapp preparing to say goodnight. He kept trying to drag his Coalition partners back to the Snowy Mountains hydroelectric scheme to connect them to a sense of nation-building, implicitly reminding the Nationals of the huge quantities of water diverted inland for their constituents' benefit. They all remained stubbornly unconvinced, longing for happier times: Monaros and World Series Cricket, *Blankety Blanks* and coal-fired power stations.

Finding firm power

Wind and solar PV are intermittent, not random. The South Australian grid operates successfully every day with around half of its generation coming from intermittent renewables. The series of South Australian blackouts in 2016 and 2017 was the result of extreme weather events, inexperience and trying to use old rules to manage new technology. It was not because operators were surprised by variances in renewable generation. Only one blackout, on 8 February 2017, was even slightly related to a weather miscalculation. On that day, the maximum temperature in Adelaide was 41 degrees and strong morning winds were delivering around 1000 megawatts (lots) of wind. This fell away sharply in the afternoon to only 100 megawatts by 6pm. The decline in wind generation was anticipated, but occurred faster than expected. When the market operator rang the Pelican Point Power Station in Adelaide and asked it to bring its mothballed second unit on line, the company advised that it hadn't been started in a year and a half and, like Grandpa's Statesman, would take four hours to restart. As a result, 90,000 households in Adelaide were blacked out.

There are three different ways to firm intermittent renewables generation: back it, store it or import it. Modern gas peaking generators are controllable and can generate large quantities of electricity quickly, and their fuel source is produced domestically.

Storing the surplus electricity from renewables includes the physical storage of pumped hydro (from Snowy 2.0 and in Tasmania), as well as batteries. Transmission imports power from nearby. Australians are already being pitched a number of potential solutions. The need to firm renewables is the big electricity policy challenge of the 21st century.

Gas

Around half of all Australian households have both gas and electricity supplied to their homes. These households recently received not one bill shock, but two. In 2016, gas prices increased sharply on the east coast of Australia, the result of a permanent, structural shift in the way gas was extracted and supplied. Australia's east-coast gas crisis started in the 1960s. Abundant, low-cost gas reserves were discovered in the Cooper Basin and then developed and supplied to Adelaide, then Sydney, then Brisbane.

The Cooper Basin gas was 'conventional'. This meant it had conveniently formed over millions of years into reservoirs around a kilometre underground, trapped in gaps in porous sandstone rock. A well-placed drill could tap into these reservoirs, and the pressure in the methane gas pushed most of it to the surface.

This made the Cooper Basin gas fields cheap to extract. In the absence of competition from overseas, state governments were able to sign up long-term contracts at sharp prices. These long-term deals then enabled gas developers such as Santos to invest in further development with confidence. Lots of cheap gas. Everyone was happy.

After 40 years, the cheap gas began to run out. The Basin had been emptied. The good news was that new 'unconventional' gas fields were being discovered throughout eastern Australia. These resources were much larger in aggregate than the Cooper Basin gas, but there was a catch: an unconventional gas field was harder to extract. The gas was often trapped in deep coal or shale seams (like gas mixed with gravel underground). These required more expensive techniques such as fracking to push the gas to the wellhead and to the surface. In other words, they were more expensive to produce.

Higher extraction costs required buyers to pay higher prices. International markets were happy to pay and export facilities were built to get the new gas to the new customers. In the end, three teams from some of the world's biggest gas businesses each developed their own Liquefied Natural Gas (LNG) export facilities in Gladstone. These are called trains because the gas goes through a number of stages, like carriages, as it is cooled to −161 degrees Celsius, then pumped into giant spherical LNG transport ships, then shipped into energy-hungry Asian markets. For the domestic gas market, which had become accustomed to a generation of low, flat gas prices, this price spike came as a shock.

The bigger concern was not that the east coast was opened to exports, but the number of LNG export facilities built. Owning export terminals was strategically important for the major gas businesses led by Santos, Origin and Shell. More LNG trains required more unconventional gas to fill more ships. In a giant game of chicken, each consortium kept building their own export facility, hoping that someone would blink and drop out. Except no one blinked. The new export contracts were, in aggregate, nearly twice as large as the entire east-coast domestic market. Development of the new gas fields accelerated in an attempt to keep up, but not as fast as was hoped. Big committed export contracts, constant domestic demand and expanding supply unable to keep up pushed eastern Australia towards gas supply shortages.

Electricity demand for gas became more volatile. As more renewables started to generate, they displaced more gas-fired electricity generation. When there was no wind and/or no solar, demand for gas spiked. This signalled the future use of gas in electricity generation in Australia: increasingly volatile demand swinging in and out even as the total quantity of gas consumed might vary very little.

Natural gas supplanted coal in the home in the postwar era because it was cleaner and more flexible: it could be used to cook spaghetti, instantaneously heat water for a shower or heat a room. It's the same when it comes to making electricity, just on a bigger scale. Gas can use large, "closed cycle" turbines to replicate the performance of coal-fired generators (though more expensively and with lower emissions). Or it can power slightly less efficient but more nimble open-cycle gas turbines to provide peaking generation. These nimbler generators are a bit like instantaneous hot-water systems, designed to quickly jump in and out to help supply spikes in demand. They are a good fit with wind and solar.

Depending on the technology mix and shape of demand, a renewable–gas hybrid will have around a quarter of the greenhouse gas emissions of an older coal-fired power station. The emissions come from two sources. Using natural gas to generate electricity emits greenhouse gases at around half the rate of coal. Natural gas is methane, which is a much more aggressive greenhouse gas than carbon dioxide if not combusted and released into the atmosphere. There are also emissions that escape when the gas is extracted from the ground. These contribute around 2.5 per cent of Australia's total greenhouse budget each year. Gas is not zero emissions by any measure. But combined with renewables it is currently the most deliverable lower-emissions electricity grid for Australia.

Because of the scale at which gas-fired generators work, they cannot simply plug into the gas mains and turn on as required. This amount of gas must be contracted in advance with gas suppliers and has to be physically delivered via a constrained network of pipelines to the power station at the time it is needed. Having enough gas to generate at specific times – or even just sustaining pipeline pressures – can be challenging, particularly in a constrained domestic gas market. The increased importance of gas in the short-to-medium term as the lead firming generator for renewables comes with its own technical challenges. It reinforces the need to get more gas into Australia's tight domestic market so that increased renewable generation can be firmed without blackout risk or electricity price shocks.

Solar thermal

During the renewables wave of the 1970s, one of the less famous classmates of wind and solar PV was solar thermal generation. Related to solar hot-water systems, solar thermal uses the concentrated heat from the sun to power conventional electricity turbines. Solar thermal technologies have evolved into dramatic-looking power stations consisting of huge troughs of mirrors, tracking disco balls, or giant central towers ringed by hundreds of mirrors focused inwards like a cult of devotees deep in solar prayer. Since the 1980s, around 20 of these fields of mirrors have been deployed, mainly in the western deserts of the US and in southern Spain. In 2004, an array of parabolic troughs was fitted to the Liddell coal-fired power station in New South Wales to preheat the water used to turn the turbines. The troughs have since been switched off.

Solar thermal was a direct competitor to solar PV. It generated at exactly the same time, so for thermal to succeed, it needed to be cheaper than converting sunlight directly into electricity. Thermal's problem was physics: it needed to make electricity the old-fashioned way, concentrating heat into steam, then into mechanical rotation, then into electricity. Starting from behind, it never caught up. There simply wasn't a business case to pay three times as much for the same renewable electricity at the same time. The last trick of the solar thermal developers was to reinvent themselves as a storage technology. If they could store the heat collected, they could generate into the evening peaks when conventional solar had gone home. Thermal couldn't compete with solar PV, but it might be able to help firm it.

The current leading technology to store this heat is molten salt, which requires high temperatures (north of 240 degrees Celsius) to stay in liquid form. Current thinking is that the configuration most likely to deliver these high temperatures is the solar tower. It's a tricky design, requiring the salt to be pumped up into a 200-metre-high tower, heated by lots of solar devotees, then stored in tanks on the ground until the heat is used for generation. The only large-scale solar thermal system with molten salt storage was commissioned in 2015 in Nevada. It has been offline or operating well below capacity for most of its life. It was closed for eight months in order to fix a leak in the molten salt tank, and its operational performance is reportedly well below what was promised. In summary, this is still an immature technology.

The technology might currently be expensive and immature, and may not actually work too well, but solar thermal with storage power stations is political gold. They're marketed as 'baseload' renewables, even though the technology isn't remotely baseload generation at all. Nor does it want to be. This peculiar definition exists because the story of a clean baseload power station replacing a dirty baseload power station is a political convenience: a highly relatable political solution to a technical problem.

Former South Australian independent Nick Xenophon first hitched his pony to the populist solar thermal bandwagon back in 2014. Two years later, he secured a $110-million federal government tax-free loan for the technology as part of the negotiations on company tax cuts. Then, in 2017, the South Australian Government signed a 'commercial in confidence' contract to buy electricity for 20 years from a new 150-megawatt, $650-million solar thermal power station at Port Augusta called 'the Aurora project'. It was less clear *how* US-based company SolarReserve won the tender, given details of the deal were never revealed and the contract price for the electricity was significantly less than what it costs to make solar thermal electricity.

The big challenge for the Aurora project appears to be raising the finance to build it. Tax-free loans and cheap government contracts are marginal in actual financial value, but they can help proponents attract prospective lenders. At the announcement in 2017, construction on the Aurora site was expected to begin early in 2018, but a year on it still has not started. Given the performance of the company's first prototype in Nevada, the risk on this sort of developmental project is material. SolarReserve has already modified its original plan to incorporate solar PV in the project, which only reinforces the real role of the molten salt system as a storage technology.

Maybe solar thermal with storage will work with enough reliability and at a price that can compete in 21st-century electricity markets. Maybe it won't. China has just built a 100-megawatt solar-thermal-with-storage power station in the Gobi Desert. Research continues on refining the technology. The issue with the Aurora proposal in South Australia is not the technology, but the politics. If governments want to back emerging technologies, their financial support should be based on a clearly defined technology strategy and bounded by transparent and competitive tenders. Hand-picking populist projects in the lead-up to an election is an indiscreet use of public monies and a poor way of discovering and developing the best new technologies.

Pumped hydro

Looking at the dark water contained behind a hydro dam is about as close as it gets to seeing the physical nature of electricity. The invisibility of electricity masks the amount of energy it contains. Energy cannot be created or destroyed; it can only be transferred. Power stations, wind farms and solar panels don't make energy. They just convert energy into electricity and then move it somewhere to perform useful tasks, from powering a laptop to carrying a bank of lifts up and down a building hundreds of times a day.

If we are going to store the huge volumes of energy required to power cities and economies, why not adapt what already works? Pumped hydro storage replicates the natural storage qualities of a hydroelectric dam. This can be done either by adapting existing hydro power stations or by creating custom-built storages. The principle is simple: stored water is pumped uphill into storage when there is surplus electricity, and discharged when extra capacity is needed.

The idea for this is not new. Pumps have been fitted to hydro dams in Europe and the US since the 1920s. There has been a pump operating at the Tumut 3 hydro power station near Canberra since the 1970s. At the start of 2017, while the South Australian Government was announcing its Tesla battery at the Hornsdale wind farm, the Turnbull government announced its bigger storage plan: Snowy 2.0. The project, reportedly first conceived when Snowy 1.0 was being completed in the 1960s, involves drilling 27 kilometres of new tunnels through the mountains and a new *Lord of the Rings*-style underground pumping and generating plant. Hydro Tasmania also has plans for up to 2500 megawatts of pumped storage across its network of dams in Tasmania. Smaller, bespoke projects are being explored in locations around the coast.

Pumped hydro storage is a proven technology. Its biggest constraint is likely to be cost. Possible sites are reportedly plentiful; commercially viable sites may be a different matter altogether. The value of the electricity stored and saved has to pay for both the capital cost of specialised pumps and tunnels and the cost of the energy needed to move tonnes of water up a hill. The critical factor for this (and batteries too) is the size of the arbitrage: pump when power is cheap, and dispatch when it is most valuable. There is a big advantage to being first into a market, because you get to buy at the lowest prices and sell at highest ones. As more storage enters the market, the effect of more buyers and sellers can flatten out this value proposition.

The Snowy 2.0 project (including new transmission lines) has been estimated to cost between $4.8 and $6.5 billion (expensive). The biggest unknown is likely to be the final cost of digging the 27 kilometres of tunnels. Some of the rock is reportedly unstable, and collapsing rock is expensive to stabilise. What this means is that Snowy 2.0 is at significant risk of costing more to build than it can ever pay back over its anticipated 50-year lifespan. If for no other reason than the political, Snowy 2.0 is likely to proceed. It may be ridiculously expensive; it may be as useful as Snowy 1.0. Such huge spending decisions should be made with cooler heads on the true benefits and costs, with reference to competing alternatives.

Physical bulk storage technologies may not be limited to using water. It's possible to store energy by compressing air in anything from tanks to underground caverns and then recovering the energy when the air is released. This type of storage has already been used in Europe and the US for more than 30 years (unrelated to renewables or climate change). The appeal of using compressed air is that it's not constrained by access to water, and, at least in theory, it could be configured to store and release large quantities of energy. To date it is one of those possible storage technology solutions that are on the 'to do' list. It is too infant to be either commercialised or abandoned. If nothing else, compressed air does remind us of the type of lateral thinking that may be required over the coming decades.

Battery

As we know, chemical batteries were invented nearly 100 years before the first electricity grids were built. The first grids used mechanical coal/steam generators, which produced industrial electricity that batteries simply couldn't match. Instead, batteries evolved an $83-billion market in delivering smaller amounts of portable electricity – car electronics, cordless drills, golf carts – as well as being scaled up to run forklifts and other light machinery. The lead acid batteries used in modern cars were developed in the mid-19th century and have remained remarkably similar for decades. Heavy and cheap, they are the dull but reliable workhorses of the battery industry.

Their eventual rivals were the 'rockstar' lithium-ion batteries, pioneered in the 1970s and first commercialised by Sony in 1991 to provide rechargeable power for portable electronics. Li-ion batteries were a revelation: lighter, more energy-dense and more expensive than their predecessors. They made possible the mobile phone and portable computing revolutions. Then Toyota scaled up another compound – nickel-metal hydride batteries – to produce its hybrid Prius car from 1997. Nickel-metal hydride was less expensive, heavier and slightly slower than Li-ion. Tesla founder Elon Musk made his intentions in the electric car market clear in 2008 when he launched the company's first electric car: a two-seater Li-ion version of the Lotus Elise that went from 0 to 100 kilometres per hour in 3.9 seconds. It was an emphatic marketing statement.

Since then, the Tesla juggernaut has gathered pace. The company listed on the stock exchange in 2010. The more money it loses, the higher the stock is valued. Tesla has doubled down on Li-ion as the core of its energy business strategy, using it to power its growing range of high-performance but expensive electric cars, home batteries and, most recently, utility-scale batteries in California and South Australia. The company has partially completed its first Gigafactory in Nevada. The facility is designed to scale up battery production and help bring down Li-ion's cost.

Then in July 2017 the South Australian Government announced that it had commissioned Tesla to build a 100-megawatt Li-ion battery attached to the Hornsdale wind farm in the state's mid-north. It was later revealed that the government contributed $40 million to sweeten the deal. The political and commercial marketing hype around the announcement was impressive – Musk even came to Adelaide to sign the deal and promised to build the new battery in 100 days, or he would supply it for free.

Currently the largest battery in the world, it is relatively small in the scale of the South Australian grid. The Tesla–Hornsdale battery can deliver 129 megawatt hours, its full capacity for around an hour. To give some sense of scale, the South Australian grid would need at least 8000 megawatt hours to run at around its minimum-demand levels over one night. That's more than 70 Tesla–Hornsdale batteries. To fully back up the South Australian grid for three cold and still days in the middle of winter would require around 1000 of them.

The battery itself has been useful and performed well. It is much faster than conventional gas generators at providing power-quality services by injecting electricity rapidly at times when this is needed to stabilise the grid. This power quality role has been the main task performed by batteries installed into high-renewable micro-grids. Claims that the battery has prevented blackouts and saved the South Australian grid are not accurate. These events simply haven't occurred. The battery wouldn't have been big enough to prevent the system black in 2016.

Musk's interest in the Australian market makes commercial sense for Tesla. As the largest global per-capita installer of rooftop solar PV, Australia is likely to be a leading market in the uptake of home batteries, if they get cheap enough. Both home battery systems and electric cars have remained stubbornly out of the reach of most consumers; the price of their Li-ion batteries has fallen, but not at the pace some were hoping for. Car manufacturers have electric car models announced and ready to go into production – once they can deliver them at key price points. Like kids waiting for Christmas, Australian solar installer businesses are eagerly tracking every flutter in the price of Tesla's Powerwall home battery and its competitors.

Li-ion home batteries and their competitor technologies still linger north of $10,000 per household (plus installation) for a modest-sized system, which in most cases still doesn't pay for itself after the 10-year warranty has expired. Australian battery sales are increasing gently to more than 6000 units a year. This is encouraging, but still a long way short of the rooftop solar PV market, which sells around 150,000 systems a year.

Li-ion is a high-performance, premium battery technology developed as a lightweight, portable, high-energy-density source of electricity for portable devices. When scaled up, these batteries deliver power quickly (the Tesla electric cars are fast), but at a cost. They use exotic and relatively expensive metals, chemicals and rare earths as key components. Most global consumption for these batteries is earmarked for cars and electronics. Li-ion batteries are finding their way into grids and some households, but the scale is still relatively low. Other simpler and heavier chemical batteries may ultimately be better suited to utility-scale use.

There are many other storage technologies under development: sodium-sulphur and advanced lead acid batteries (called ultrabatteries). Flow batteries are a different design of battery that uses two different chemicals dissolved in liquids, often separated by a membrane where the charge is released. Most of these are not quite as fast or as energy-dense as Li-ion, and they're heavier. But they are less constrained by expensive inputs and therefore may be easier to scale up to provide bulk power.

Chemical storage doesn't need to be contained by batteries. Surplus electricity from renewables can be used to electrolyse water to produce hydrogen. That hydrogen gas can then be stored for later reconversion back into electricity via a fuel cell, burned to produce heat or shandied with natural gas as a combustion fuel. The idea of 'a hydrogen economy' has been around since the 1970s: it's the idea of using hydrogen as another energy vector, an alternative storage system that can be produced by zero-emissions technologies and then used to power transport or generators, or for heating.

The early promise and broad potential of hydrogen makes it an ideal political backstop for low-carbon energy policy makers. In January 2019, Labor leader Bill Shorten announced a $1-billion National Hydrogen Plan, including an 'innovation hub' to be built in the Queensland port of Gladstone, usefully located in a key marginal electorate. Like its competitors, hydrogen will ultimately need to store and move energy effectively and at a competitive price to make it a competitive option.

Transmission

Another way of firming increased renewables penetration is to increase interconnection to other grids, and then import the power required. This is part of the logic behind the decision of the current South Australian Government (led by Premier Steven Marshall) to throw $200 million at a new interconnector between South Australia and New South Wales. Connecting state electricity grids was an important step in creating the National Electricity Market (NEM), allowing states to trade electricity, lowering costs and increasing reliability. During the recent heat waves, the network of transmission lines was working overtime shifting electricity into high-demand states. South Australia benefited significantly from its interconnection to Victoria in the pre-renewables 1990s, as the much larger, cheaper coal-based Victorian grid generally always had plenty of surplus electricity to sell into South Australia's more expensive market.

Transmission will play an important role in a transforming electricity sector, but it doesn't generate any power: it only moves it. If you are going to connect to another state to help back up your renewables, it's probably a good idea to make sure they will have enough surplus power to help you out when needed.

Capture

The idea of capturing carbon dioxide and injecting it underground was originally developed as a resource-recovery technology by the oil and gas industry long before climate change became an issue. Carbon dioxide was often found in underground natural gas reservoirs, and gas companies needed to separate it from the valuable methane at the wellhead. By the 1970s, industries started reinjecting this waste carbon dioxide back into active oil wells, pushing more oil to the surface and boosting output. Usefully, under the right geological conditions, the carbon dioxide tended to stay underground. The first oil and gas demonstrations of carbon capture and storage (CCS) were simple adaptations of this proven and commercial technique.

It was a much bigger technical leap to try to adapt this process, bolting it onto power stations to capture their greenhouse gas emissions. Most coal-fired power stations were not near active gas or oil wells, and the bolt-on of what amounted to a large chemical plant to siphon off the carbon dioxide from their smoke stacks, compress it and pump it underground used around a third of the energy produced by the power stations. It quickly became obvious it was uneconomic to retrofit this technology to older coal combustion technologies, which effectively ruled it out for Australia's existing fleet of coal generators. Even the adaptation of the technology for new coal generation has left investors underwhelmed in terms of cost and performance.

Despite early interest from the coal industry and talk of a multi-million-dollar voluntary fund to support the technology, most of the research investment has come from governments. Significantly, in 2017, when the mining industry entered the political debate of energy policy, they proposed the building of relatively conventional HELE (high-efficiency, low-emissions) coal generation, not CCS. HELE technology uses a slightly more efficient combustion of coal, which slightly reduces emissions intensity through improved performance, but at a higher cost. It's like buying the latest premium model of the same car.

More efficient coal generation will not solve the emerging challenge of how inflexible baseload generators integrate with intermittent generation at scale. As South Australia discovered, renewables at scale will eventually push all other generators out of the market for periods of high wind and/or sunlight. As the renewables dial back, they require generation or storage that can quickly re-enter the market. This is not how CCS, HELE or conventional coal works.

Meanwhile, global CCS research has continued. Its biggest drivers are oil and gas companies who are exploring ways to incorporate the carbon dioxide released by gas-fired power stations into the production process. Their latest technology, called the Allam Cycle, is currently being trialled in a 50-megawatt demonstration project in Texas. It creates and then recycles high-pressure carbon dioxide, rather than steam, to turn the power station turbines. Eventually, the waste CCS is compressed and pumped underground. It still faces the same basic questions as other technologies: how much does it cost, what are the full emissions from the production cycle and where would the carbon dioxide go? CCS doesn't appear likely to resuscitate coal-fired electricity in Australia any time soon, but it may yet be adapted to reduce emissions in gas generation and other industrial processes.

Nuclear

Australians have four points of reference when they think about nuclear power: Three Mile Island, Chernobyl, Fukushima and *The Simpsons*. None of them are very positive. What most Australians don't realise is that we started building a nuclear power station in 1969. The Gorton government thought it would be a good idea to build a (small) 500-megawatt nuclear power station to produce electricity and provide weapons-grade plutonium for the development of nuclear weapons in Australia. The site chosen was on Commonwealth territory, on the southern side of Jervis Bay (200 kilometres south of Sydney), just near the secluded Murrays Beach. Tenders were issued, the site was cleared and concrete was poured. A change of prime minister halted the project after Treasury warned of the costs involved. Today, the site is used as the car park for the nearby beach.

Australia's participation to date in the use of nuclear energy has been confined to the export of uranium to a select group of nuclear-energy-generating countries including the US, Japan, Korea and France. Debate on the issue remains muted. Active public discussion of nuclear energy remains confined to a handful of enthusiasts on Twitter. Self-evidently, there is a strong political and moral debate that tends to overwhelm any discussion of its technical properties.

Nuclear energy was developed as an integrated part of the proliferation of nuclear weapons during the Cold War. Activist environment groups such as Greenpeace were formed protesting nuclear testing, first in Alaska, then the Pacific. In part because of this association, and in part because of concerns about the environmental risks and hazardous waste produced, the environment movement has historically regarded nuclear energy as a non-negotiable. This absolutist position has been challenged with the emerging understanding of climate change and the proposition that nuclear is a possible zero-emissions energy solution.

Nuclear energy's biggest problem in the political debate is that its failures have been so substantive and so tangible. There remain large exclusion zones around the sites of the nuclear accidents at Chernobyl and Fukushima. The uneven radiation from the Chernobyl explosion in 1986 means the site itself will not be safe for 20,000 years, while surrounding areas will take between decades and several hundred years. Wildlife has since returned to the exclusion zone and is flourishing, while curious nuclear tourists are allowed into parts of the abandoned post-apocalyptic town of Pripyat.

As you would expect, current nuclear technology is much safer than the type built at Chernobyl between 1977 and 1983. The Fukushima nuclear power plant was commissioned in 1971, the first in Japan. It was hit by both an earthquake and tidal waves in 2011. These disabled its cooling systems, resulting in radioactive releases and the establishment of a 30-kilometre exclusion zone around the site, which remains today.

After the accident, almost all of Japan's other nuclear reactors were switched off. Due to sustained public anxiety they remain off. The Japanese government is now looking to build new coal-fired power stations to replace its nuclear capacity. In short, if faced with the option of climate change or nuclear power, the Japanese people would prefer climate change.

The continued widespread use of nuclear energy, and the broad antipathy towards it, creates a kind of mass cognitive dissonance on nuclear. According to the International Atomic Energy Agency there are still 453 reactors operating in the world, 99 of these for more than 40 years. Another 55 reactors are under construction, mainly in China, Russia and India. Anyone who has travelled in Europe will have effectively used nuclear-generated electricity, or possibly driven near one of more than 100 reactors operating there. The French government is currently building a major underground nuclear waste storage facility in the Champagne region. Countries in Europe that have banned nuclear reactors are still happy to import the electricity they produce.

From an operational perspective, a nuclear reactor works much like a 'baseload' coal-fired power station, except the heat source is a reactor rather than a coal boiler, and they don't produce greenhouse emissions. This baseload inflexibility makes it unsuitable as a technology to firm large-scale intermittent generation. New iterations of nuclear are designed to be safer, smaller and more flexible, but they have very high capital costs resulting in some of the most expensive electricity on the market – around $300 per megawatt hour according to the recent CSIRO GenCost projections of generation costs.

The high cost of building new nuclear is reflected by the debate surrounding the Hinkley Point C nuclear power station project in the UK. The UK currently sources around 15 per cent of its electricity from nuclear power stations. Like coal-fired power stations in Australia, these are reaching the end of their working lives and need to be replaced. The planned Hinkley Point C is massive – around 3200 megawatts, or seven per cent of total UK demand – and current estimations suggest it will cost at least $36 billion and take ten years to build. Planning for the project began in 2010 but construction has still not started. The UK government is currently reviewing the project, having agreed to underwrite it with ongoing payments to ensure it remains competitive in future electricity markets, which could dramatically increase the costs even further.

Thorium reactors use a more abundant fuel than uranium and the reactors themselves are designed to avoid meltdowns. On top of this, the fuel cannot be used to produce weapons-grade plutonium. (Thorium was originally overlooked as a nuclear fuel because it couldn't be weaponised.) The less controversial, more science-fiction relative of nuclear energy is fusion energy. It's a theoretical technology that seeks to extract the energy of fusing atoms together rather than splitting them. Fusion energy has been under development for decades but is still in the early stages. The joke in the energy industry is that fusion energy will always be 30 years away. For the purpose of the current energy policy debate in Australia, it is not remotely on the radar.

Nuclear-generated electricity is greenhouse-emissions free, but it will be expensive to install and take up to a decade to establish the regulatory regime required for their operation, safe waste disposal and transport. These up-front costs would only be warranted if multiple nuclear generators were built in Australia. It's an all-or-nothing type of energy strategy.

In the current highly charged political debate around electricity in Australia, a rational discussion of nuclear energy is simply not feasible. If it does emerge, it will most likely be a few years from now. The majority of Australians are probably more comfortable seeing how far renewable energy integration can go first. Perhaps if higher levels of renewables do become technically problematic, then nuclear energy may re-emerge into the future energy debate.

What's the future for firm generation?

As markets mature and costs become more competitive, the range of technologies involved tends to narrow. Renewable energy had a wide field of potential technologies in the late 20th century. Today wind and solar PV have left the rest for dead. The new technology contest is around the most efficient way to firm wind and solar PV's intermittency. In contrast to the consolidation evident in renewables, the developmental state of the firming technology market is illustrated by the number of possible solutions: pumped hydro, compressed air storage, flow batteries, hydrogen, solar thermal, super capacitors, flywheels and a catalogue of chemical batteries made of lithium-ion, ultra-lead acid, nickel cadmium or sodium sulphur. As development continues and the field narrows, gas can provide flexible backup as renewables set the pace for new generation. A national energy policy design should ensure that we get enough generation when it's required.

It's highly desirable from an emissions and efficiency perspective to be able to store surplus generation from intermittent renewables for use when they're not generating. The value case for providing these storage services depends largely on their access to cheap electricity, most likely surplus renewables generation. One of the critical challenges for market re-design will be how to enable firming technologies to compete and develop while delivering efficient and reliable outcomes for electricity customers.

Coal power's ability to reposition itself as a credible lower-emissions option by using storage technology doesn't appear to be going anywhere. This is reflected in the evolving siege mentality from the coal miners: they are now the world's biggest exporter of coal in an economy that isn't building any more coal. They continue to enjoy high international prices and sustained demand from Asia, including from many emerging economies. New coal-fired electricity generation appears increasingly unlikely in Australia. Low-emissions gas may extend its shelf life if new turbine technology proves successful. Otherwise it will be displaced by whatever emerges from the contest between storage technologies.

And in the background is nuclear power. Expensive, uneasy, inflexible but dispatchable zero-emissions. Rationally, the threat posed by climate change is greater than the risk posed by nuclear energy. If it came to it, what would Australians do?

7

WHY IS THERE SO MUCH ROOFTOP SOLAR IN AUSTRALIA?

The remarkable story about solar PV in Australia, the thing that sets us apart from any other country in the world, is the extraordinary boom in rooftop household systems. At the start of the 21st century, Australians were watching TV reports about solar as a futuristic energy technology. Ten years later they were watching TV ads spruiking it to them. Australians in their thousands then bought and installed rooftop solar PV. By 2010, solar companies were sponsoring Australian sporting teams. Solar PV for your home became as normal a consumer purchase as buying a dishwasher. Solar panels have saved Australian households thousands of dollars, whether through overly generous subsidies or, more recently, the falling cost of the systems.

Australians' embrace of solar panels as a consumer product was a complete global outlier. It was a policy accident transformed into a multi-billion-dollar consumer market by a combination of political expedience and sustained falls in solar technology cost. The rest of the world used solar panels to build power-station-sized arrays. This was the logical and cheaper way to deploy solar PV at scale. Household solar is more popular but less efficient: it takes more equipment to produce the same amount of electricity; some of the panels get shaded by trees or other buildings. More critically, as household solar has scaled up, it has imposed additional costs on (or is constrained by) suburban networks not designed to move these volumes of electricity in the other direction. Rooftop solar didn't go crazy because it became super-efficient – it went crazy because governments went crazy for solar.

The household solar PV numbers are staggering. In 2007, around 8000 Australian households had rooftop solar PV. Now there are more than two million. That's 20 per cent of the entire Australian residential market, including flats and apartments. Australia has easily the highest rate of household solar in the world. The penetration rates in some of the most solar-dense suburbs of Brisbane and Adelaide are around 70 per cent. That's two out of every three houses.

In all, there were four key factors in the second half of the 2000s that created Australia's unique residential solar boom. One, residential solar PV subsidies became a populist political promise from governments trying to dig themselves out of trouble at a time of drought and anxiety about climate change. Two, while governments were doing this, the cost of the panels fell dramatically because of big changes in the supply chain and global economy. Three, Australia's urban sprawl was highly suited to rooftop solar (lots of roof space and lots of sun). Four, all these factors occurred during a period of sharply rising electricity prices and heightened consumer awareness about ways to bring these down.

A double subsidy

As discussed earlier, the Howard government started subsidising household solar PV in 1999 almost by accident. At the time they were negotiating with the Australian Democrats to secure support for the national Goods and Services Tax (GST). As part of the final deal, Howard agreed to a package of mostly token environmental reforms, including a household solar PV subsidy. The Photovoltaic Rebate Program (PVRP) was launched on the first day of the new millennium, initially providing up to $8250 to any household that installed an accredited solar PV system. This might seem overly generous now, but back then the entry-level price tag for solar was in excess of $15,000 per house. The original idea was that the PVRP scheme would trickle through a few hundred household solar systems until the $10 million set aside for it had run out, and then it would close. Within its first year, the subsidy was dialled back to $7500, and then to $4000 in 2003 as solar PV systems got cheaper.

Separate to the PVRP, Howard's MRET paid renewable energy generators, large and small, for the amount of power they produced. Small-scale technologies such as solar PV were eligible, and to make things easier, the value of the generation was estimated over the life of the system and then paid up front (deeming). This added about another $1000 to the existing household solar PV subsidy. Despite the introduction of two separate subsidies, rooftop solar PV uptake was initially modest. In the first year of the MRET only 118 solar PV panels were fitted to Australian houses. By mid-decade, household solar PV was shifting at a modest rate of around 1000 systems a year.

But at least household solar was doing better than the utility-scale technology. The MRET scheme subsidised the cheapest utility-scale renewables, and at the start of the 21st century, solar PV was relatively expensive compared to wind farms or bioenergy projects such as landfill gas or bagasse (sugarcane waste). So while in other parts of the world solar PV was primarily a utility-scale technology, the only solar PV being installed in Australia was onto homes. Australians logically assumed that solar PV was just another household energy technology, a bit like solar hot-water systems.

Sticking a solar hot-water system – a hot-water tank and a black solar collector – on one's roof is a simple but effective technology that has been around since the 1970s. For most of the noughties, solar hot water was 10 to 30 times more popular than rooftop solar PV. The reason was simple: a solar hot-water system after the MRET subsidies cost around $2500. A solar PV system cost a minimum of at least $10,000. Over the 10-year life of a solar hot-water system, the savings from lower power bills would cover the extra cost of the solar hot-water system. The negative payback for solar PV limited the market to enthusiasts.

In the second half of 2006, the debate on climate change turned on a dime and ripped the Howard government's political head off. Howard scrambled furiously to try and stem the political bleeding before the November 2007 election. He grabbed the PVRP scheme and turbocharged it. In his pre-election budget, Howard doubled solar subsidies through the PVRP to $8000 a household – a double serving of middle-class welfare. It was the sort of highly political but poorly conceived policy announcement that soon became commonplace in energy policy.

Over the first decade of the 21st century, the cost of solar had been falling steadily with incremental improvements in panel technology and economies of scale in the supply chain. Getting an $8000-cheque in addition to the $1000-subsidy from the MRET (the subsidy depended on the size of the rooftop solar system and its location) sent the value proposition for solar PV into overdrive. The

PVRP went from a budget cost of a few million a year to eventually costing taxpayers more than $1 billion. Household installation of solar PV took off.

Howard lost the November 2007 election, but his rooftop solar PV subsidies hit multiple hot buttons with the public. Solar was clever, clean and green, it reduced electricity bills and it created a sense of energy independence for households. Consumers liked to own the solution themselves. Middle-class home owners could generate their own power and didn't need to rely on power companies. Solar PV had a vicarious popularity: people liked other people owning it, not just themselves. Putting solar on suburban rooftops may not have been the most efficient way of deploying the technology, but households didn't know that, and they didn't care. Solar was tangible. You couldn't see an emissions trading scheme, but you could see, and touch, rooftop solar. It felt like something was happening.

Within a year, state governments were climbing aboard the solar-subsidy bandwagon. They couldn't afford to throw cash at middle-class Australian solar buyers, but they could tweak the way solar households were billed for electricity. They introduced solar feed-in tariffs that paid households a premium for every watt of solar electricity sold back into the grid. The Victorian Government kicked things off with a 28-cent-per-kilowatt-hour feed-in tariff in 2008 (which was cranked up to 60 cents in 2009). In other words, households that installed solar panels were getting paid for their solar electricity up to three times what it cost to buy it. Governments in South Australia, Queensland and New South Wales followed suit, siphoning money from non-solar households and using it to pay generous deals to their solar neighbours. Those who couldn't afford solar, or who didn't own their house, were paying for those who did.

It should have been seen as an outrageous political rort, a manifestly inequitable transfer from the have-nots to the haves. Instead, it was wildly popular.

By the start of 2009, many eligible buyers of solar PV qualified for three separate government subsidies. The number of household solar installations increased tenfold within a year. Solar in Australia became one of, if not *the*, most heavily subsidised consumer products in history.

The power of cheap technology

While governments were throwing money at households to install solar, the system costs kept falling. This was driven by three factors: rapid escalation in the scale of Chinese panel manufacturing, structural changes in the cost of the main material used in solar PV – silicon – which brought down costs, and a glut of panels caused by a downturn in global demand in the wake of the global financial crisis. Solar PV costs started to tumble just as government kickbacks reached their peak. The combination of these two factors sent household PV into the stratosphere.

The global PV market had grown steadily through the start of the decade behind renewable energy targets in Europe and some US states. These targets drove utility-scale solar farms, with most panels being supplied from Europe, Japan and the US. Then, in 2005, China's Renewable Energy Law happened. With China dominating the global solar PV market, the brands of solar panels sold to Australians changed rapidly: in 2005 you were offered Kyocera, Sharp, BP or Conergy panels. By 2010, the leading brands were Suntech, Trina and Canadian (yes, Canadian Solar is Chinese).

Around 2008, the source and type of silicon used in panels also became cheaper. Until then, the solar PV industry had managed to source most of its silicon from semiconductor scrap. But continued growth in the size of the solar PV market meant this *Steptoe and Son* approach had become a constraint, and prices for the scrap silicon began to skyrocket from around US$30 a kilogram in 2004 to US$475 a kilo in 2008. Development began on a new solar-grade silicon that could be made to a less expensive specification without compromising electricity output.

When the global financial crisis hit, Chinese manufacturers simply slashed prices to shift volumes. One of the few solar markets that were picking up was the Australian residential market. Discounted Chinese panels arrived by the container load.

Boom

Companies spend millions of dollars on market research and advertising agencies, trying to find out how consumers think and what motivates the purchases they make. Finding ways of encouraging consumers to buy your product is a multi-billion-dollar global business. In the emerging consumer market for solar PV, none of this happened. There was no market research, no focus groups. One day Australian consumers barely understood what solar PV was, and the next they were diving into their wallets, ready to spend a few thousand dollars. It was an instant mass market. This shift in consumer attitudes blindsided pretty much everyone – the solar industry, governments, and many of the major energy companies.

One of the ways of determining whether a technology such as solar is a worthwhile investment (after the subsidies) is to estimate the payback period: how long does it take for the savings on electricity bills to cover the up-front payment? The general rule is that it makes no sense to buy something if it takes longer for the device to pay itself off than its expected life. With solar, the payback rule was pretty simple: anything longer than 10 years was too long and not worth it. But many households didn't do the maths. They just bought solar when the entry-level price fell inside the range of a large credit card purchase, around $3000. Under that price point consumers felt empowered to buy solar on the same basis that they made most other purchases: because they liked it. It's likely that other highly desired consumer energy products such as electric cars and batteries also have their own price point. If and when this is reached, consumers will leap into the market.

As the price of solar was falling, electricity prices started to go up. The cost of running the electricity network in New South Wales and Queensland went up after major blackouts in those states. The state governments' responses were to 'gold plate' – or overspend – on network reliability. On top of this, one of the impacts of the Millennium drought was to limit some coal and hydro generation capacity because of water shortages. The resulting higher demand and reduced output further pushed up power costs.

By 2008, consumers could see their power bills going up and were keen to embrace a technology that could immunise them from bill shock. This sense of energy independence has been a recurring theme in the Australian household solar story. Many consumers talk of 'going off-grid' when they buy solar, even though they are very much still on the grid, and probably using it more than before (because they both export and import power). Some convinced themselves they would be spared from blackouts (they aren't) and some even managed to convince themselves that their rooftop solar system worked at night (it doesn't).

Solar PV generates emissions-free electricity. The climate-change anxiety unleashed by severe water shortages and rationing in Australia's major cities in 2006 blindsided and destroyed the Howard government. For Australians, installing rooftop solar was a visible and tangible way of reducing emissions, and that contributed to the solar boom. Installing solar saved money. The systems could be bought over the phone with a credit card, they were cool, high-tech and hassle-free, and a tangible way to reduce greenhouse gas emissions and reduce reliance on energy companies. It felt good in so many ways. Put together, these consumer sentiments unlocked a new, white-hot $5-billion residential solar PV market.

Who was buying it?

The election of the Rudd government in 2007 might have been a logical time to adopt a more responsible approach to solar PV. Hardly. The popularity of solar was contagious. Despite the flagrant nature of Howard's desperate solar cash handout, Rudd did not unwind it. Instead, he simply restricted eligibility for the $8000 cheques to households earning under $100,000 a year. As it turned out, most of the households buying solar were already earning less than $100,000 a year.

The big buyers of solar were electricity price-sensitive home owners: cash-strapped young home buyers and retirees (or those approaching retirement). Solar sales were much softer in affluent suburbs, although the design of the net feed-in tariffs (where you get paid for what you export back into the grid) made solar popular with owners of beach houses. In 2012, the top solar suburbs were in retiree/beach suburbs such as Caloundra in Queensland and Victor Harbor, Goolwa and Aldinga in South Australia, or in cash-strapped outer mortgage suburbs such as Ellenbrook and Pinjarra in Perth,

Hallett Cove in Adelaide and Jimboomba near Brisbane. Toorak and Mosman were noticeably absent from the list.

Accelerating this trend was the curious relationship that emerged between solar PV and mandatory superannuation. In 1992, the Keating government had introduced mandatory superannuation payments by employers. This was, in effect, to institutionalise retirement savings for all workers. By 2010, those approaching retirement had around 20 years of these modest super contributions in their funds. Many retirees had superannuation balances that, combined with other assets outside the family home, were at or slightly over the maximum limit set (around $258,000 for a home-owning couple) to qualify for the full age pension. The further they were over this limit, the more their pension payments would be reduced.

Spending down surplus cash by putting solar panels on their house transferred liable assets to non-liable assets, while reducing or eliminating a recurrent cost (energy bills). Financial planners commonly advised their retiring clients to install solar PV, and tens of thousands of retirees took up this sensible option. The size of these systems was often determined by how big their roof was.

Who were the winners?

The Australian household solar boom had some obvious winners. Solar households that bought at the peak of the subsidies and premium feed-in tariffs got free money for new technology that cut their power bills and emissions and made them feel good about both. Depending on the size and location of their system, the up-front investment sometimes paid for itself in a year or two. Another big winner was the fledgling solar industry. At the peak of the subsidies boom, selling solar panels was like shooting fish in a barrel. Self-proclaimed entrepreneurs were created overnight. The industry's favourite sales pitch was 'Buy before the government subsidies run out', which should have been a warning about the durability of their business models. When the subsidies did end, so did many of these businesses.

The high churn within the fledgling solar industry had little impact on customers, in part thanks to rigorous safety and accreditation requirements set up by the Howard government. Rooftop solar systems had to be registered, installed by an accredited electrician and signed off with a certificate of safety. It was a hidden blessing. The solar boom followed on the coat tails of the subsidies for household insulation – the now infamous (and lethal) pink batts scheme. In 2010, it was discovered that around 20 per cent of household solar systems in Australia had a circuit-breaker switch incorrectly installed, which put them at (minor) risk of sparking and possibly starting a fire, if repeatedly turned on and off. The faulty installs were repaired quickly by the installers within weeks.

The other big winners from the residential solar boom were governments. Solar went from being a minor part of a political deal on the GST to a headline-grabbing gut political play. Governments of both colours could see the political upside in being pro-solar. Labor governments justified their solar subsidies as tangible action on climate change. Coalition governments cast solar as a way of helping battling families cut their power bills. Their political instincts were validated: once the postcode data was made available by the regulator, it revealed much of the solar was going into marginal electorates. The subsidies delivered cash into the laps of swinging voters who could throw governments in and out of office.

The rank political opportunism of solar policy masked its

poor design. There were virtually no controls on the location of the systems. Solar was installed where people bought it. At first small numbers of systems made little difference, but as the numbers increased, so did their impact on parts of the electricity network. The main system problem was high penetration of solar PV in parts of the grid, which pushed up the local voltage levels (the electricity pressure in the network). The role of the substations scattered throughout the grid is to drop the higher transmission voltages down to around 230 volts. The power is then pumped to the surrounding customers. Near the substation the pressure is normally a little higher, and by the time the power gets to the houses furthest away from the substation, the pressure may sometimes be a little lower, say 220 volts.

A household solar PV system, when it is exporting, puts small amounts of power back into the local network. A single system will have a negligible impact on voltage, but high densities of solar in some suburban streets started to push the voltage higher. This tends to be most extreme at times where there is low demand but high sunlight, such as noon on a winter's day. Maintaining over-high voltages is neither safe nor good for electrical appliances. The high voltages also caused some rooftop solar systems to switch themselves off when the pressure in the street became too high for the rooftop solar system to push its electricity onto. This meant those solar customers weren't getting full value from their system and the inverter (the box that converts the DC from the solar into AC in the grid) was at risk of failing early because of constant tripping.

It will cost the networks many millions of dollars to upgrade the grid to accommodate the extra solar, which would in turn be paid for by all consumers: another transfer from non-solar to solar households. Welfare groups have quietly become concerned about the growing inequity of continued solar subsidy and feed-in tariffs. The cost of small-scale solar subsidies became the biggest part of the renewables bill in most states. At its most inequitable in 2015 in Queensland, the cost of funding the state-based solar feed-in tariff was around $270 a year for every household. That's a lot, when you don't have much.

The wind back

By the middle of 2009 Australians were buying more than 100,000 solar panels a year – costing the Rudd government more than $800 million a year. Rudd knew that this was unsustainable. He needed to unwind the increasingly expensive and over-generous support mechanisms without killing the industry or infuriating the public. In June that year he announced the sudden end of the $8000 subsidy and replaced it with increased support for solar under the Renewable Energy Target (RET), which would then be dialled down gradually as solar costs fell.

State governments facing elections responded by increasing their feed-in tariffs. Victoria's increased to 60 cents, then New South Wales increased their feed-in tariff to 60 cents a kilowatt hour using gross metering, which meant the solar household got paid for every watt generated from the system, even if the household used the power themselves. The New South Wales Keneally government's reckless over-generosity was the high-water mark in the race to over-compensate solar. The wind back of this deal (by the new O'Farrell government) led the gradual deflation of solar-subsidy politics in Australia. Solar sales increased as subsidies declined: 200,000 in 2010, 360,000 in 2011, 343,000 in 2012. Sales have stabilised at around 150,000 systems a year since then, surging to more than 200,000 new systems in 2018 off the back of rising electricity prices, increased commercial installs and still-falling system costs. After lagging behind solar hot water and heat pumps for a decade, solar PV now outsells them by a factor of two to one. Solar PV has become a permanent feature of the residential energy landscape in Australia.

Game changer

For the past three decades solar has continued its relentless march down the cost curve. Continued efficiency gains and competition between different panel technologies have seen solar fall below US$1 per watt, which is below US$60 per megawatt. As costs kept falling, global growth in solar PV has been exponential.

Because solar PV generation is intermittent on a daily basis and demand is not, it is optimised if paired with appropriately sized storage. The cost of household-scale storage remains the main barrier to its integration. More solar households with storage would help avoid over-voltage in high-density solar PV suburbs and help flatten demand peaks. Governments just subsidising storage would effectively amount to yet another cash transfer to the same cohort who have been beneficiaries of a decade of over-generous government funding. The leading value of storage technology is in power-quality services. To deliver this value, having networks of household solar systems, each supported by a battery and controlled remotely – a virtual power plant – makes more sense. These can be brokered by a third party: a retailer, network or just an independent aggregator. This approach may allow the greater benefits of storage to be realised faster, as well as accelerating their cost-effective deploy.

But as if we haven't learned the lessons of the past, the political fashion of middle-class cash handouts is making a comeback. The Victorian Government has introduced a solar PV (and hot water) rebate of up to $2225 for eligible households. The Labor federal Opposition is promising $2000 rebates for household batteries. This won't solve the supply–cost barriers holding batteries back and it won't direct the batteries to where they might do the most good in the network. Those who don't own their home or who live in apartments are locked out of the government handouts, and therefore, directly or indirectly, end up paying for them. Again.

Australia's solar story is both a lesson and a warning for other economies. Rooftop solar is now a growth market in parts of the US and Europe, while utility-scale solar is now rolling out in Australia. Befitting a country that sits on the other side of the world, Australia got its solar story completely arse-up compared to everyone else. We over-subsidised expensive rooftop solar and ignored cheaper utility solar. In doing so we were first to discover both the benefits and challenges of a more distributed grid.

8

CAN RENEWABLES GRIDS EVER BE RELIABLE?

Transitioning from the thermal stability of the 20th-century grid to one increasingly dominated by intermittent renewables is a radical step. It would make a lot of sense to carefully plan and model this transition, think through what might go wrong and put in place measures to mitigate these risks. But we didn't do any of that in Australia. Instead we began to transform the biggest machine in the country – the electricity system – with a design based on nothing more than political compromises. When a carefully designed carbon trading scheme floundered in parliament, the compromise solution was to install 5000 wind turbines, or other renewable technologies with the same output, and see what happened. There was no specific mechanism to steer the location of these new generators or give the market operator greater flexibility to manage the disruption. We just let it rip. The Australian way.

Unexpectedly, nearly half of this renewable investment piled into South Australia, which turned it by default into an accidental experiment in large-scale renewables integration. The people of South Australia became lab rats, going about their daily activities, turning electrical things on and off, not realising that, a bit like *The Truman Show*, they were living inside a live, full-scale electricity laboratory – a grid being unwittingly stress-tested until something went wrong.

The test results following the blackout in September 2016 were mixed. On the plus side, a high-renewables grid was technically possible and mostly reliable. Day to day the new renewable generators worked as specified, their output could be predicted and the grid could successfully adjust. The fails were on the market side. Without some sort of overarching policy design, high renewables penetration undermined the commercial viability of the firm generators needed to support them. The existing market design also made South Australia over-reliant on importing lots of cheap brown-coal electricity from Victoria. This had the effect of inflaming the impact of disruptions that occurred immediately after the transmission line was upgraded. Having too little South Australian firm generation running on 28 September was the trigger for the statewide blackout. These were valuable, albeit painful, lessons.

Armed with this information, the South Australian grid has been running without incident. The suite of new political investments has, to date, been useful but had a marginal impact on reliability. Overall, Australia is now about a third of the way to integrating renewables into a large, stringy, isolated electricity system. More testing and experimentation is needed, but preferably not on South Australians if we can help it. If only there was a way of building scale models of the electricity grid so we could conduct further trials and testing without losing another state's power system.

As it turns out, there is.

An archipelago of grids

Australians like to romanticise our wide brown land, our sunburnt country, but the truth is that most of us live in big cities at the edges of it. Nine out of ten Australians live in sprawling coastal cities and towns along the south-eastern corner of the continent. Or Perth. The electricity grids that supply power reflect this. The National Electricity Market (NEM) hugs the coast for the entire 5000 kilometres between Port Lincoln in South Australia and Port Douglas in Queensland. It's the world's longest interconnected electricity system. A second, smaller grid radiates around Perth like a wagon wheel.

Outside of these two main grids, electricity is supplied by an archipelago of more than a thousand small, islanded systems and micro-grids. Like a set of Russian dolls, these grids range in size from the North-West Interconnected System in the Pilbara, then the Darwin–Katherine grid, then Mt Isa, Alice Springs, Broome, and down to smaller islands like Coober Pedy in South Australia and King Island in the Bass Strait. The smallest of these are hundreds of remote communities and isolated cattle stations.

The cost of supplying electricity to these remote locations is significantly higher than inside large urban grids. This is because supplying gas and diesel fuel to these locations is expensive, and because the grids are less efficient due to their smaller size. It costs Horizon Power, which supplies power in remote Western

Australia, around $4000 a year more to supply a remote household than one in Perth. This difference is paid for by state government subsidies.

That makes these grids ideal places to integrate renewable generation. They are more competitive at reducing expensive fuel costs, and can potentially help resolve some of the technical challenges and limitations of a high-penetration renewable electricity system.

The King Island Renewable Energy Integration Project

King Island sits in the middle of the Bass Strait, almost due south of Melbourne. It's a windswept island known for its cheese factory and not much else. The island's electricity is supplied by Hydro Tasmania, which in the 20th century ran four diesel generators to supply power to around 1600 residents and the cheese factory. When small commercial wind turbines became available in the late 1990s, Hydro Tasmania decided to trial the new technology on the island's windy east coast. Three small turbines were installed in 1998 and worked in synch with the diesel system via a regulator. As the wind increased, the electricity produced by the turbines increased and the diesel would automatically feather back to maintain stable voltage and frequency. It burned less diesel too. The savings in diesel fuel inspired Hydro Tasmania to investigate further. How much could renewables do? Could they ultimately power the entire island?

To increase the supply of renewable electricity, two much larger wind turbines were added to the mini-grid in 2003 along with a developmental type of storage: a vanadium flow battery. It's a popular misconception that grid-scale batteries (including the 'world's largest' Tesla battery in South Australia) work like a mobile phone battery: recharge when the wind is blowing and then power the grid for hours when the renewables have stopped. This is stretching the truth somewhat. Large-scale batteries can and do sometimes store and discharge, but for minutes, not hours. The Tesla battery can discharge around 100 megawatts for an hour, or 50 megawatts for two hours, or 25 megawatts for four. This is in a state that is typically using around 1000 to 2000 megawatts most of the time.

Grid-scale batteries are currently nowhere near running a grid, even for a minute. But they are particularly effective in providing power-quality services, and working alongside high penetrations of renewables to stabilise power fluctuations, like filling in potholes on a road. Batteries are excellent at providing short bursts of power for seconds or minutes, to smooth things out when the wind dips briefly, as it often does in windy conditions, or when clouds roll over on sunny days. They make a high-renewable grid smoother and more reliable. The supply of bulk backup electricity on King Island still comes from diesel. Not the battery.

After the vanadium flow battery was installed, trial and error ensued. The battery was developmental technology at the time and failed soon after installation. Hydro Tasmania soon replaced it with a bank of dynamic resistors designed to soak up the surplus electricity from the bigger wind farm. The extra wind generation could regularly produce more power than the island grid could handle. In this simple configuration King Island could only cope with about 40 per cent of its generation coming from wind. If the turbines exceeded this, the surplus electricity was just blown off by the resistors as waste heat, like filling a glass with water till it spilled over the sides.

The reason King Island 2.0 couldn't handle any more wind power was to do with those two critical features of an electricity grid: voltage and frequency. These are the elements that define the quality of power. The voltage can vary a bit around 230 volts without too many problems. If there is too much pressure, then your appliances are at risk of burning out faster. As is regularly depicted in science-fiction movies, voltage surges would cause old incandescent light bulbs to glow more and more brightly until they blew. Low voltage would cause the same lights to dim and flicker.

Frequency is related but different. Frequency is like a musical note that hums at the exact same pitch in every wire of your house and across the grid. This 'note' is the electromagnetic wave that is created by the spinning motion of a generator. The rotation of the alternator inside a generator causes the current to flip backwards and forwards with each half-rotation. That's why it's called alternating current, and that's why these generators are called alternators. This note is held at a constant rate of 50 cycles per second (50 hertz), and it has to be completely uniform right across the grid. Every electrical device in Australia is configured to work in tune with this 50-hertz note. Even modest changes in frequency risk can cause serious damage to large and small electrical devices and assets across the grid.

When the frequency of the grid is disrupted and cannot be reigned back in quickly, the grid shuts sections down to protect itself and keep the system safe – this is how we get security blackouts (which include the system black in South Australia in 2016). Sudden frequency changes can occur when a generator trips off, or when power lines are suddenly brought down in a storm. At these moments, it's absolutely critical that frequency is restored within a few seconds. To do this, you need something big and powerful that is either already operating or can be switched on quickly to lean in and push the note back to 50 hertz. When there are a lot of big generators operating, they can combine to push the frequency back to normal. Large chemical batteries have also proven to be excellent at this job, and some fast-start gas can do this as well. It's a big, powerful note across the system, and it needs a big, powerful response to push it back.

Imagine an all-male Welsh choir from the 1970s: beefy lads, ruddy faces, ill-fitting suits, rugby front-row thick necks. This is like Australia's fleet of 20th-century coal-fired electricity generators. Setting frequency in the grid is like asking them all to sing a single note in unison. One of the Welshmen starts to sing the note and the others join in. As a choir, it's easy to sing and maintain the note. When one takes a breath or a rest, there are plenty of other voices to keep singing. If something suddenly happens to one of the singers, the others can lean in, sing a bit louder and cover the missing voice.

Solar and wind are like a jazz saxophonist, playing with the choir. They play in and out of the note, around it and sometimes on it. Sometimes they play and sometimes they don't. They *can* play the note, but they can't hold it. When there are lots of Welshmen singing, keeping the note going is not a problem. But if they start to leave, you need to find someone or something else that can do the job. Managing frequency is one of the big technical challenges for high-renewables electricity grids, large and small.

On King Island, the same rules of voltage and frequency applied as everywhere else. Power quality was relatively easy to manage in the pre-renewables grid. Governors attached to the diesel generators throttled the output and kept voltage and frequency nicely in balance. As the wind generation increased, Hydro Tasmania found that they still needed to keep plenty of diesel operating to control the grid. Even if the wind was blowing strongly, they could only let wind account for about 40 per cent of total generation, otherwise it became too unstable.

Given that King Island is quite windy, this meant that a lot of perfectly useful wind energy was being mopped up as waste heat in the bank of resistors, rather than being sent to King Islanders to make cheese or boil a kettle. To get renewables to run deeper into the grid would require more technology. In 2010, after careful planning and design, Hydro Tasmania rolled out the King Island Renewable Energy Integration Project (KIREIP). The objective was to see if they could run the island on 100 per cent renewable generation/'diesel-off' status for as long as possible. To achieve this, three new technologies were installed: first, the failed vanadium redox battery was replaced by a more reliable lead acid battery; second, a bank of solar PV panels was added, mainly to see how wind and solar integrated into the system; and finally, two specially designed flywheels were developed to provide the Welsh choir that would be needed when, and if, the diesel generators could be turned off.

A flywheel is a heavy spinning mass that physically absorbs and releases energy. It takes a lot of energy to get going, and a lot of energy to stop. Flywheels replicate the rotating mass, the inertia, of a thermal generator. In the case of King Island, once spinning, the heavy wheel would replicate this inertia. If the frequency of the grid started to deviate from the 50-hertz note, the flywheel's energy would be enough to step in, haul it back to normal, and then step out again. The flywheels had a small diesel engine attached so that, if the forces driving frequency deviation were sustained, the small diesel could quickly start up and power the flywheel to maintain inertia in the grid while the full diesels were readied to be brought back into service. The new lead acid battery would also help with power quality, discharging quickly in variable conditions to keep power quality in check. Combined, these technologies meant King Island should be able to run using pure renewables.

Once the new system was bedded down, the wind and new solar renewables were able to power much more of the island than before. As planned, in strong, sustained winds, the diesel generators switched off completely. The King Island micro-grid has managed to run for nearly two windy days straight using only renewable energy. Once the wind dies down, the system switches back to diesel generation. In total, renewables now supply 65 per cent of the island's electricity. While Hydro Tasmania is coy about the total cost, the 2010 integration project cost around $18 million, or more than $11,000 per King Islander. The cost of the original wind turbines and resistors is additional to that.

King Island is a controlled electricity experiment, done with low risk to supply: the full diesel generation system was available as backup in the event of any system faults or fails. There were teething problems, and reliability was affected in the initial stages, but this has improved over time as operational knowledge has improved. King Island has provided valuable information on what electricity grids need to keep working in the thin altitudes of high-renewables generation. It serves as a reminder that bulk power supply is still needed when renewables are not generating. Bulk power supply is not a battery. At least not yet.

Most importantly, King Island's success was due to its ability to conduct its research without political interference. There were no politicians announcing new batteries or solar thermal power stations on the island. Perhaps it's a useful reminder to leave the grid design to the engineers, and the emissions targets to the politicians.

What's even better is that you can watch it work from your phone. There is an app that provides live data on how the King Island grid is working at any time (search 'KIREIP').

To Hydro Tasmania it's more than just an experiment. The system at King Island was a business model that has subsequently been configured into a shipping-container modular system and is now being rolled out in other remote grids. The KIREIP modular system was first tested in neighbouring Flinders Island, then Coober Pedy in remote South Australia and Rottnest Island in Western Australia. It's now regarded as one of the leading exponents of small grid-scale renewables integration in the world.

Alice Springs

In 2004, as the climate change cold war was escalating, the Howard government announced a $100-million Solar Cities program. Criticised at the time as a tokenistic fig leaf, it reflected the government's growing recognition that the public was embracing this new technology. Solar Cities was sold as a demonstration program for solar to fund towns and communities to install solar technologies and learn from how they worked. The political reality appeared to be a bit more self-serving: many of the Solar Cities projects ended up in marginal electorates.

In the lead-up to the 2007 federal election, Alice Springs was announced as a Solar City. It seemed an obvious choice. Alice Springs is iconic in Australian culture: a small, extremely sunny town slap-bang in the centre of Australia. It was a crowd-pleaser. For this isolated community of 27,000 people, Solar Cities was a big deal. And why not? Their sunlight was genuinely world class. Alice Springs receives more than 300 clear days a year, some of the best uninterrupted sunshine in the world. The $42-million program supported installation of residential and commercial solar, solar hot-water systems, energy efficiency, smart meters and new tariffs. As much as anything, it was something to unite the community. After decades of living under all that sunshine, they finally got the chance to do something useful with it.

Remote grids such as the one in Alice Springs are not only expensive, they're slightly less reliable too. Technical problems have to be solved without the ability to import power, and with many of these grids scattered through northern Australia, they're regularly exposed to more extreme weather events. When Solar Cities was announced, Alice was powered by an ageing gas and diesel power station. There wasn't much technical planning for how the new solar generation would integrate into this micro-grid. As elsewhere, the main technical focus was on the safe and reliable installation of the rooftop solar systems themselves.

From 2008, solar panels started to be installed on top of hotels, at the airport, in a field south of the town and on Alice Springs houses. The Territory Government offered an extra-generous feed-in tariff to solar households to push things along. There was a real solar buzz in Alice Springs. They hosted renewables solar conferences and a solar-research facility was set up to test panel performance.

Unrelated to the new solar boom, a new replacement gas-fired power station was being built out of town. Changing over power stations can cause some temporary reliability issues, particularly in a small grid: the reconfiguration can reveal new constraints in the poles and wires, and new generators can trip more frequently.

In the second half of 2010, blackouts started to occur, increasing in number through 2011. There were seven blackouts over that financial year, including two system blacks. The levels of solar PV at that stage were modest – around two megawatts in a grid with an average demand of around 25 megawatts. The main culprits appeared to be the old retiring generator and some network issues arising with the relocation of the new generator. It was assumed that the new generators would improve the reliability of the system. Just to be sure, the old generator was kept in service. So there was a lot more backup generation if needed. But after this transition was bedded down, the blackouts didn't go away. They became more frequent.

What was going on? The local utility noticed that the frequency of the grid was becoming less stable on those rare cloudy days. When a cloud came over, the solar generation fell rapidly, then powered up again as soon as the cloud passed. At low levels of solar PV this wasn't an issue, but it would soon become a bigger problem if the scale of PV was allowed to increase unchecked. They also observed that when a fault occurred in the network or a generator tripped – which pulled frequency away from the 50-hertz note – the first thing to happen was that the inverters in the solar PV systems tripped off. This meant that the solar households and hotels went from electricity exporters to importers. In other words, instead of leaning in as conventional generators did during a frequency event, the rooftop solar systems leaned back and inadvertently made things worse.

An added difficulty was the lack of visibility of the solar PV. The panels were scattered throughout the grid, and couldn't be controlled or monitored. They worked like a domestic appliance (which was what they were) and operated according to their own settings. In a small grid, the increased solar PV added a new type of volatility. Small faults or disturbances became bigger ones. Grids regularly experience minor events that are corrected by automatic systems. Cutting off power to customers was very much a last-resort solution. But it was happening with concerning regularity.

Other similar remote grids with high levels of solar PV had also seen these problems coming. In 2010, Horizon Power in northern Western Australia put a moratorium on the installation of new rooftop solar PV in towns including Broome and Carnarvon once they reached around 10 per cent of generation. It wasn't a very popular decision, but it gave them time to work out how to manage small remote grids with high solar penetrations. There were no moratoriums in Alice Springs. As the Territory Government encouraged more systems into its flagship Solar City, solar experts gave the system the all clear. A one-megawatt solar farm was commissioned in 2011 and expanded to four megawatts in 2015. Alice Springs had 10 megawatts of solar PV by 2015 and 14 megawatts by 2017, often reaching 30–40 per cent of total generation.

In an attempt to reduce the emerging risks, the new generator, Territory Generation, increased the spinning reserve (the gas and diesel generators powered up and spinning in reserve in case of a fault). Running high-spinning reserve is expensive – it burns fuel for generators that aren't producing electricity. On cloudy days, they tried to turn down solar as much as they could to minimise frequency fluctuations, which were getting bigger as more solar PV piled in.

In January 2016, another Alice Springs system black occurred in the middle of a heat wave, snapping the Territory Government into action. The Alice grid had been flying blind as it tried to manage high-penetration solar PV. It probably needed a King Island makeover – or, at the very least, a battery. Instead it got new gas and diesel generators. Only in 2018, following a change of government, was a five-megawatt battery installed.

The problems haven't gone away. With no check on rooftop solar approvals, Alice Springs is heading towards 100-per-cent renewable power at times of minimum demand (cool, sunny spring and autumn days) by as early as 2022. Like in King Island, this will technically force the gas and diesel generators off. Unlike in King Island, there is still no plan for how to manage this. King Island has a renewables micro-grid designed by engineers. Alice Springs has a renewables micro-grid designed by press release. It's an important difference.

The issue here is not the use of solar PV, which is a good idea in a sunny, remote micro-grid such as Alice Springs. It's the lack of planning and the reactive way decisions were made. Eight months after the 2016 system black in Alice Springs, South Australia had a full-grid version of the same thing, with many of the same root causes. The Utilities Commission of the Northern territory noted in its 2016–17 Power System Review that 'if left to grow unmanaged, solar generation will detrimentally affect the secure operation of the power system'.

Alice Springs will remain a live test site for renewables integration. From it, we can continue to learn a lot about how solar PV is different to wind at a high level of grid penetration; how the electricity from small, distributed generators moves across the network; and how to optimise its performance.

King Island and Alice Springs are at the extremes of the experiment. Most renewables integration trials in remote grids have followed a sensible, risk-managed approach. Ergon Energy used a small solar farm to displace around eight per cent of its diesel generation in the remote Far North Queensland town of Doomadgee. Horizon Power in northern Western Australia capped solar in its remote solar towns, but has also experimented with early installation of grid batteries to optimise solar, and has used varying incentives to steer more rooftop solar PV into locations where the grid is weaker.

These technical challenges are not the exclusive domain of small grids. Large remote grids such as the South West Interconnected System (SWIS) around Perth are heading in the same direction, and in a similar timeframe. Around 25 per cent of Perth households have already installed solar PV, and the rate of new installations is accelerating. If these rates continue, then the solar PV in Perth's larger but isolated grid could be supplying 100 per cent of that grid's minimum demand by 2025. Logically, it's reasonable to assume the SWIS will need something like the same suite of supporting technologies that are deployed in King Island, only a thousand times bigger.

Mining

The real test of all these experiments, modules and microsystems is whether they can deliver reliable, cost-effective off-grid electricity in the real world. And it doesn't get any more real than the mining industry. Resource extraction consumes about 10 per cent of Australia's total energy. Mining is a logistics business on a massive scale. It's about moving millions of tonnes of earth and rock to extract minerals and ores that are then processed and sent to customers. Mining operations go wherever the resources are, frequently locating in some of the most remote and sweaty parts of the continent. Given these distances and scales, there is little time for niceties. The cost of operating in these harsh and remote environments places a premium on reliability and durability. Risks of all types must be minimised, which means miners are more reluctant first movers on emerging, less proven technologies. They are, in this regard, the corporate antithesis of renewable-energy developers.

When governments and households became enthralled with wind and solar at the end of the 2000s, the miners couldn't have been less interested. They were in the middle of a ferocious resources boom. Every fibre of their corporate beings was bent towards getting hold of the trucks, equipment and people needed to extract resources while their price boomed. It was only in the post-boom cycle that their focus turned back towards greater efficiencies and cost savings. When the price of solar and wind began to tumble, it became more feasible for them to consider using renewables to offset the cost of remote gas and diesel supply.

But there were still risks. The hybrid renewable systems were more complex and less reliable than standalone diesel. A solar PV system would have a working life of 10 to 15 years, so it needed to be matched to a mine that was going to be operational for at least that long to justify the investment. Then there was risk around the projected savings. The price of diesel and gas is linked to the world oil price. Chronic oil price volatility made it hard to do reliable estimates of how long renewables might take to recover their capital costs. And then there was first-mover risk. Why had no one else done this? Why should they go first? Mining companies could see renewables integration being installed into remote mining towns such as Marble Bar and Nullagine, but they knew those grids were heavily subsidised.

The first 'brave' miner to move was Rio Tinto, induced by $3.5 million of free government funding in 2014 to install solar panels at its bauxite mine at Weipa, near the tip of Cape York. It was a logical fit. Rio Tinto already had experience with electricity, running grids in Western Australia's north-west and Far North Queensland. As high-profile global businesses, both Rio Tinto and BHP were under increasing scrutiny for what they owned and how they invested. Rio Tinto had begun quietly divesting its coal-mining assets from 2015. The 1.7-megawatt Weipa solar PV facility displaced diesel burned at the 26-megawatt generator. It was a low-risk, entry-level renewables integration that didn't compromise reliability. A small toe in the water.

Next, and considerably braver, came junior copper miner Sandfire, which built a $40-million solar PV and battery system at their DeGrussa mine, 900 kilometres north of Perth. Eleven megawatts of solar PV and a six-megawatt lithium-ion battery were bolted onto the mine's existing 19-megawatt diesel generator, cutting consumption by around five million litres a year. To help reduce risk and induce more miners to the idea of integrating renewables, more than half of the cost of the project was funded by the federal government. With the subsidy, and depending on the fluctuating diesel price, the project should pay for itself in around four to five years. Without it, it's about break even.

Solar and wind costs have continued to fall, but this hasn't triggered a renewable off-grid mining boom. However, a trickle of projects has been announced. In 2018, BHP offshoot South32 completed its installation of three megawatts of redeployable solar at its Cannington mine in Queensland. Another three-megawatt solar installation was announced for a mineral sands project north of Perth. Copper miner OZ Minerals is looking at a range of options, including hybrid renewables, to power its Prominent Hill mine in the north of South Australia after it was kicked off the existing transmission line by BHP, which wants to use all of its capacity to expand another mine. Hybrid renewable generation is still nowhere near being the go-to technology for these applications; concerns about reliability risks still outweigh the marginal savings.

What can we take from this?

Some of the first locations to install renewable generation in Australia were off-grid. Remote generators continued to explore and experiment with renewables as they looked for ways to reduce the high cost of providing reliable, remote power. In the process, they have developed the suite of technologies needed to operate high and even 100-per-cent renewable generation for sustained periods. As it turns out, this is particularly handy to know now, given some of Australia's major towns and cities (Perth, Adelaide, Alice Springs) are heading towards minimum-demand events supplied entirely by renewables generation from early next decade.

These various off-grid experiments around Australia have not only informed the technical requirements needed to coordinate wind and solar at scale, but they have also demonstrated the critical role of planning and coordination of all new generation, both utility- and household-scale. It would be useful to continue with the experiments that have already started: to see what practical differences exist between high-wind and high-solar generation, to refine and improve reliability, and to optimise the scale and operation of the new supporting technologies that will be needed such as synchronous condensers and batteries.

It might be useful to explore the different ways of monitoring, controlling and supporting high levels of distributed solar PV to maintain power quality at the lowest cost. It would appear that Alice Springs is an ideal location to explore these options before they are scaled up in real time and tested in the real world, when Perth begins running on pure solar PV around the mid-2020s.

King Island is a world-leading micro-grid in the integration of intermittent renewable energy, and it has managed to reach 65 per cent renewables. Clearly the bumper sticker promise of 100-per-cent renewable energy by a certain date is currently outside the technical envelope – it's like promising crewed flights to Mars. In 2013 the AEMO modelled the feasibility of reaching 100-per-cent renewable energy in the NEM. This research was one of the conditions of the Greens deal with Julia Gillard in 2011. The AEMO found that a 100-per-cent renewable power system would cost at least $219 billion, but probably more. The future scenarios modelled by the AEMO relied heavily on geothermal, wave and concentrated solar thermal technology to deliver firm generation, as well as a large expansion of biomass. This is no longer a credible premise. While there have been big advances in storage technologies, they are nowhere near providing the bulk power capacity needed to replace either the diesel generators in King Island, or their equivalents across the NEM.

9
WHAT CAN CONSUMERS DO?

For the decades up until 2007, most Australians' quarterly electricity bills were unremarkable and predictable. Then everything changed. For the next decade electricity prices increased in real terms by 56 per cent – much faster than incomes – while other comparable services stayed much the same or got cheaper. This was caused by a combination of unrelated factors: first, after a series of blackouts in New South Wales and Queensland, overzealous new reliability standards from state governments pushed a spike in network charges. Then came the short-lived carbon price and its repeal. Then the closure of the Northern and Hazelwood power stations reduced electricity supply, spiking wholesale prices. This was exacerbated by high gas prices and tight coal supply in New South Wales.

These explanations had little resonance with consumers, who couldn't see any justification for the cost of electricity going up and up and up: the lights didn't shine any brighter; the more expensive electricity didn't do anything useful like provide wi-fi or stream movies. There were two basic reactions: consumers looked for ways to bring bills down themselves, and the growing outrage energised escalating political intervention.

Households consume around 30 per cent of electricity in Australia. The rest is used in commercial buildings and industry. Industrial customers, particularly large electricity users, also became alarmed at the material shift in one of their input costs. Electricity as a share of household spending was only around two per cent, but any cost increase was amplified against a decade of flat real wages growth. This underlying sense of economic stasis for consumers translated into the political concern with 'cost of living pressures'. People didn't feel like they were getting anywhere, and governments empathised with them, vowing to get 'cost of living pressures' back down. The only problem was it was hard to actually deliver. Real wages were flat; so too were the real costs of most goods and services. The spike in electricity and gas prices stuck out. They became a lightning rod for a much broader sense of public dissatisfaction. And so they became a relentless target for political attention.

The 'cost of living pressures' debate ran parallel to the increasingly politicised debate about climate change. While related, the two issues stoically ignored each other like feuding neighbours. Over time, this price–climate fault line has widened, the left prioritising climate and trying to ignore the cost of promising big targets, while conservatives focused on prices, mostly trying to avoid acknowledging climate change at all. Technologies such as household solar PV, which speak to both narratives, have benefited from being promoted on both sides of the divide, an accidental outbreak of bipartisan support.

Of course, transforming the electricity system will need to address *both* emissions and price. The role of the electricity consumer is already changing in the 21st-century grid. Until a decade ago, electricity consumers were passive; they just sat at the end of the pipe and paid their bills. Consumers are now playing a more active role: generating, shifting their demand and even starting to lean in and help support the grid when things get tight. Households and businesses are already becoming solar micro-generators. A grid that was designed around distributing electricity from large power stations is now trying to cope with a more distributed system. How will this new grid operate and what do consumers need to think about as they become more active players in the grid?

Falling demand

The simplest way for consumers to reduce electricity bills and greenhouse gas emissions is to use less electricity. This can be achieved by behaviour change (turning lights off, setting air conditioners to more moderate temperatures) and technology change (more efficient appliances and automated energy-saving devices). The two biggest sources of residential electricity demand in Australia are heating water and heating and cooling air. This water–air combo on average consumes around two thirds of a household's electricity. The rest is used for refrigeration, pool pumps, cooking, powering kitchen appliances, washing and drying, lighting and then a tail of other electrical devices including televisions and computers. Averages mask big variations in household demand. Air conditioners run up much bigger power bills in Darwin than in Hobart. Big houses cost more to heat or cool. Appliances such as clothes dryers are big electricity consumers if used regularly. Managing this requires little more than common sense. Everyone knows the drill.

The real challenge with encouraging greater efficiency is that no one really consumes electricity; they consume the appliances that it powers. No one slips a record onto their turntable and ponders how many kilowatt hours it might consume that night. We watch television, dry a load of washing, have a shower and turn on the air conditioner because it makes life better and easier. Big power bills, if charged quarterly, can come weeks after periods of heavy electricity use (such as summer heat waves). By then, the connection between the purchase and the cost is weak. This fragile relationship poses a conundrum for policy reform. Most customers would rather not think about electricity at all. Plenty would probably prefer a 'fixed-price all-you-can-eat' power bill like in a mobile phone or internet plan. But this would do nothing, on its own, to drive greater efficiency.

Reflecting this, the political response to energy efficiency has been largely symbolic. This has included populist adaptations of schemes (such as the Greenhouse Gas Abatement Scheme in New South Wales in 2007) that ended up giving away low-energy light bulbs (the similarly tokenistic Victorian Energy Efficiency Target, or VEET 1.0 scheme that was in place in Victoria until 2014). These schemes just performed the political theatre of energy efficiency, indiscriminately handing out retail consumer products that many households would have just bought anyway.

Independent of these mostly hopeless efforts, per-capita electricity consumption has been falling steadily in Australia since 2009, after half a century of electricity growth that faithfully tracked GDP. The sudden and surprising break from economic tradition was driven by a range of factors: big chunks of falling demand from deindustrialising sectors such as car manufacturing and aluminium smelting; rising prices from 2007 driving a reaction from consumers; the rarely acknowledged benefits of increasing household insulation through the ill-fated pink batts scheme; and the continued efficiency gains of new electrical appliances. (New air conditioners can be around 30 per cent more efficient than those installed 10 years ago.) Most of these efficiency gains have been passive: energy savings driven by natural replacement and upgrade.

Hot-water systems

Too many people make the decision about their next hot-water system when they are naked in a suddenly cold shower at 7am, having just discovered that the old one has died. What happens then is an emergency purchase – often a like-for-like replacement within 24 hours to get life back to normal as soon as possible. It's a missed opportunity. Rising energy costs increase the value of efficient water-heating systems. This is a mature market with fierce competition between major brands and technology types: solar hot water boosted by electricity or gas, electric heat pumps and instantaneous gas.

Choosing an efficient hot-water system is a considered purchase. Different models of heat pumps and solar hot-water systems are configured for different climates. Solar hot-water systems are made up of roof-mounted collectors that use sunlight to heat water. This is then stored in a tank on the roof or on the ground. Evacuated tube systems (where the tubes are made up of two layers of glass with vacuum between them, increasing insulation) tend to be more expensive and more efficient. Their main rival is the heat pump, a ground-mounted tank that uses the energy in the ambient air by running it through a heat exchanger, like an air conditioner in reverse. It's solar energy, just in a different form.

There has been an ongoing debate within the hot-water industry about which technology is better. Solar hot water is a more passive technology that heats during the day and requires a sunny, unshaded roof. Heat pumps are more versatile, but rely on a small amount of energy to run the fan and heat exchanger. Leading solar hot-water and heat-pump systems are both highly efficient, and recover their higher up-front costs within a few years. A well-planned hot-water system will make a measurable difference to household electricity costs.

Your electricity bill

The way Australians are billed for electricity is a legacy from the government-owned power-station era of the mid-1970s, before the rise and rise of household air conditioners and rooftop solar PV. We are still billed for electricity as if it's water coming out of a tap. In practice, the cost of supplying electricity surges and retreats like a tide. The process is dynamic. There are periods of cheap abundance followed by periods of scarcity and higher cost. And yet the way we are billed smooths these out. This didn't matter in a largely coal-fired grid. It matters much more as we move to more intermittent generation.

The first bill design problem is that households are incorrectly charged for grid access: being connected to the physical network of poles and wires. The real cost of network access is typically somewhere between 30 and 50 per cent of total electricity costs. Most bills charge around half this – somewhere around a dollar a day, or $30 a month, or $365 a year. The rest is smeared into the unit cost of electricity bought. This discrepancy has become more important in the past decade because the current method ends up undercharging solar PV households, even though they are still big users of the network, exporting and importing electricity. Given this is about paying for the grid to be maintained and connected to your house, solar households should pay the same as everyone else. This is about fairness. When two million households are under-charged, by default everyone else has to pay more.

Historically, the electricity system has been like a devoted servant: whatever customers asked for, they got. There has been no constraint, and no limit on how much power consumers could use or when they could use it. As demand continued to increase, new generators were commissioned and networks were upgraded – alternative solutions were not even considered. The billing system reflected these times.

Now that those days of constant, cheap electricity are over, a more intermittent 21st-century grid is going to need more flexibility from consumers: to help reduce demand when electricity supply is tight, and take advantage of times of abundance. The way we pay for electricity will be the trigger. This means both charging what it actually costs to use the network, and then starting to signal the varying high and low costs of electricity at different times. Of course, that increased complexity can't just be pushed onto consumers without some help. Since most electricity consumers don't think about the electricity when they're using it, they're unlikely to be effective at managing dynamic tariffs on their own. Without the right technical support, dynamic tariffs would be political suicide. It's unrealistic and verging on irresponsible to expect customers to monitor constantly changing prices before deciding whether to put on a load of washing or watch TV.

Instead, using more dynamic tariffs would need to be part of the shift to what is called a 'smart grid'. There are already more than

80,000 households around Brisbane that have discounted network charges in exchange for direct load control being installed on their air conditioners. This enables the network provider to remotely take control of the thermostat during peak demand days. The networks can also remotely control more than a million hot-water tanks and more than 35,000 pool pumps to ensure they switch off during demand peaks. 'Demand response' is a related measure, where subscribing businesses and households get paid an agreed fee to automatically reduce their load during peak times. Load shifting takes this even further, using dynamic tariffs combined with advanced metering and communications systems to enable the network provider, retailer, market operator or the house itself to reduce load during peak times and fill up on cheap electricity at times of abundant generation.

The technologies for these smart grid systems have already been trialled in Australia. Some retailers have trialled offering more dynamic tariffs to their customers, but take-up remains low. It's a chicken-and-egg problem: more cost-reflective tariffs on their own are only attractive to a very small proportion of households who are willing to take an active and constant interest in managing their consumption, or those whose demand spikes already fall outside the conventional peaks (such as shift workers). So there is no real change in behaviour and no benefit to the system.

Dynamic tariffs require new, smarter technology to do the switching for the customer. And the value for the investment in smart electricity technology comes from the savings made using dynamic tariffs. It's an evolving process: these types of tariffs are more likely to support the financial case for installing home-storage technologies by increasing the opportunities and value of the arbitrage between battery charge and discharge. There are other (more expensive) versions of conventional technologies around that only make commercial sense if they can optimise dynamic tariffs, such as ice-storing air-conditioning technologies for larger buildings (where ice is made overnight and used to cool air during the peaks).

Some current retail deals do include a peak vs off-peak rate, but these are still relatively analogue in their approach, charging a higher rate between 7am and 11pm and a cheaper rate overnight – a coal off-peak rate. These try to encourage greater consumption of electricity at night; when demand falls away, coal-fired generators cycle down to their lowest settings and wholesale prices are traditionally cheapest. This is why many conventional electric hot-water systems were set up to switch on in the middle of the night. In South Australia, with high penetration of solar PV, the new off-peak time is starting to move to mild and sunny afternoons. The minimum-demand moment in the South Australian calendar year is now around 1.30pm on Boxing Day.

This is why the new buzz words in demand management are 'energy flexibility'. Being able to shift demand to different times of the day will be a key feature of the 21st-century grid. Ideally it will end up producing even simpler bills: households paying a flat rate every month, as with internet or mobile phone plans, and the supporting technology working diligently behind the scenes, massaging how each house and business works. This kind of electricity management will require round-the-clock diligence. It will require coordinated automation, which will require a suite of other technologies to enable this flexibility. Starting with a smart meter.

Smart meters

If the way we are billed for our electricity use hails from a time 50 years ago, the way more than half of Australian households still measure electricity comes from the 19th century. The 'traditional' spinning metal disc inside old analogue electricity meters is a primitive technology in a digital age, like using a typewriter instead of a computer. These meters have to be manually read every three months, with all the cost and access issues (locked gates, dogs) that this entails. The new generation of digital smart meters is an essential upgrade in the 21st century. They more accurately measure usage throughout the day and can report this data back to customers in real time via an app or website. This connects the electricity consumer to the act of consumption.

Smart meters were infamously installed across Victoria from 2009 in a mandatory rollout, an excellent initiative poorly executed. The rollout invoked a hostile reaction from some, who objected to the tiny electromagnetic transmitters that sent the data back to the network provider and retailer. Their concerns related to the supposed health impacts and/or the privacy impacts of electricity companies collecting more detailed information on their consumption. Anecdotal reports out of the US described the ability of the police to find and arrest hydroponic dope growers based on their unique electricity demand patterns.

Given the meters were installed outside and the electromagnetic waves were less than those produced by a baby monitor, it's more likely the mandatory rollout itself was the problem. Some people may oppose being given free iPads if they were forced to take one. With hindsight these concerns weren't well handled, and the Victorian network providers could have defused the problem if they'd been allowed to let individual households opt out of the rollout, so long as they covered the full cost of the manual meter reads.

Smart meters enable consumers to more accurately track electricity use, get more frequent billing and help networks identify faults instantly, which drastically reduces outage time. In the longer game, smart meters are critical enablers of more efficient and dynamic use of electricity, helping to match demand with supply and keeping power bills down in the process. They are the critical interface that changes a building from a passive electricity receiver into a dynamic part of the machine itself.

Shopping around

The 1990s shifted electricity in Australia from a government department to a competitive market. This process also introduced retail competition. Competitive retail markets remain the simplest and most effective way of ensuring that consumers are not being over-charged for a good or service, whether it's marketplace fruit and vegetables, airline tickets, groceries or electricity.

But there is a catch. The effectiveness of competitive markets depends on there being sufficient and active engagement by consumers. If you don't use it, you lose it. Sellers of goods and services across capitalist economies have evolved complex strategies to protect and maximise margins inside this constraint. They do this by competing fiercely on price to win customers and then over time, where possible, rely on customer lethargy (some call it loyalty) to recover increased margins back from these mature customers. This is standard operating procedure for a whole range of services that are consumed constantly but purchased infrequently, such as insurance, mortgages and energy.

Its antidote is simple but slightly fatiguing: it pays to shop around. It doesn't help if consumers look to change retailer every few months, as the cost of this retail 'churn' (constant quitting and re-signing) can become expensive for all customers. But it does pay every few years for consumers to test the market. In electricity, the Australian Energy Regulator (the third of the major national institutions along with AEMO and AEMC) runs an independent comparison website called energymadeeasy.gov.au to assist consumers in finding the best deals in the market for their particular usage. This involves about 20 minutes of entering bill information into the comparator site, getting the range of best deals and then either making the switch, or getting this price matched by your existing retailer.

When electricity retail competition was introduced, it was expected to drive innovation in areas that government-run utilities weren't particularly interested in: more choice in billing and deals, better customer service and the use of retail technology. After 20 years, retail electricity markets are price competitive, but the expected reforms have been underwhelming. For an industry grappling with radical technology changes in generation, its marketing techniques are still defined by the telemarketing and door-to-door sales of the 1950s.

For most consumers, electricity remains a grudge purchase. Most don't really care about innovations, they just don't want to pay too much. Retail practices have reflected this. Some innovation in the way electricity is billed has come from smaller retailers. Even with sustained increases in electricity costs, consumers have remained lethargic. This customer lethargy is one of the reasons smaller electricity retailers persist with telephone marketing and door to door sales techniques – as annoying as they may be, they remain one of the most effective ways of finding and recruiting disinterested consumers.

One area of innovation that has long been speculated about is whether electricity bills would ever be aggregated with other services, such as telephony or insurance – or whether large retail businesses such as Coles or Woolworths would enter the market. There have been a handful of attempts to bundle telephony and electricity, but nothing yet to suggest that this is likely to be a natural evolution of the sector. Major supermarkets and telephone businesses have kicked the tyres about getting into electricity, but to date they have held off. That's presumably because the risks are bigger than the margins. The chronic uncertainty of the past decade may also be stifling this type of commercial innovation.

The lethargy observed in electricity may also reflect a broader issue of consumer fatigue. The number of retail relationships maintained by consumers has increased continuously over the 21st century. A generation ago, a household had to manage only a handful of 'accounts' that led them to communicate with businesses (bank, electricity, rates, telephone landline, video library) compared to the dozens they have today (internet, mobile phones, Netflix, Google, YouTube, Facebook, app-based parking meters, tolls etc…). Besieged by multiple businesses wanting to build and manage a relationship, maybe most consumers are just over it by the time they come to buy something as uninteresting as electricity.

Competitive retail electricity markets are likely to be much more important as the grid transforms. Different households and businesses will have varying appetites to shift their load, and varying ability to install new technology and other mechanisms that will elicit greater demand response. The way to discover these different consumer patterns is through competitive markets. Like a Soviet disco, unitary one-size-fits-all markets simply miss the point: it's what you *didn't* think of that can be the most intriguing discovery. Maybe working with customers through this next transformation will be what drives innovation in electricity retail and what ultimately creates a more successful relationship for customers.

Building

In the 20th century, people came in their millions to live in Australia. It might have been a bit far from everything but the weather was nice and there was lots of space. The quarter-acre blocks and spacious houses of 20th-century urban development celebrated this. And so Australian cities grew to gargantuan sizes – physically: greater Melbourne today is nearly 10,000 square kilometres, with a population of 4.5 million people. Greater London houses nearly 8.6 million in 1570 square kilometres. Throughout this growth, energy efficiency was rarely a priority for builders or owners, and those years of cheap electricity, along with a temperate climate, masked the energy-efficiency failings of Australia's urban sprawl.

Annual per-capita electricity consumption by Australian households ranks around the middle of the pack when compared with other developed countries. That's because while many of us live in leaky and inefficient houses, we only tend to heat and cool them for a few months each year. This makes our electricity demand peakier: demand surges in heat waves and cold snaps, but flatlines in spring and autumn. It also keeps a lid on total electricity costs. Australians still only spend 2.17 per cent of their average disposable income on electricity, up from 1.99 per cent in 2011. (This is only the average though – the share of electricity costs are higher for low-income and vulnerable households.)

The upside of this stock of large, energy-inefficient houses is that many Australian households have plenty of room to fit rooftop solar PV. The downside is that retrofitting these buildings to improve energy performance is mostly prohibitively expensive. After installing ceiling insulation and low-energy light bulbs, the rest gets harder: double-glazed windows reduce heat transfer, but cost thousands; wall insulation is more expensive to install than ceiling insulation; draughts can be isolated and blocked, older air conditioners and heaters can be replaced. New building standards in some states do require improved performance, but most new houses continue to be large and therefore remain energy-intensive to heat and cool.

The legacy of energy-inefficient housing is felt most acutely by the 30 per cent of households who rent. They are stuck with whatever they signed a lease for. Renters, both residential and commercial, remain largely forgotten in the rollout of middle-class welfare for new solar panels, batteries and hot-water systems, as landlords have weak incentives to improve the energy performance of their rental asset unless they can realise this in higher rents – which leaves renters right where they started.

Attempts to make the rental property market more transparent when it comes to the energy performance of buildings have been challenging. The Australian Capital Territory has had a mandatory energy efficiency rating (EER) scheme since 1997, which requires the landlord to disclose the EER if one has been calculated. Poor-performing buildings simply aren't rated or disclosed. Accurate methodologies are complex, and therefore audits are expensive. Successfully tackling the split incentives of rented residential and commercial buildings is a problem that has bedevilled policy-makers for more than a decade.

The solar revolution

It's worth emphasising again that two million Australian households now have solar PV installed – nearly the same number as have subscribed to Foxtel. Solar PV is now just another consumer product: an option for Australian home owners to manage their energy costs. And as with Foxtel, prospective consumers want to know if it's worth the money, and how it will work for them.

The basic pre-condition for anyone interested in buying solar PV technology is that they need to own an unshaded roof. Where exactly the panels are placed is open to debate. The installers of the first generation of solar PV systems received generous premium tariffs that created a financial incentive to maximise output and export generation. As a result, solar panels were installed facing north, to capture the most sunlight all year round. As these premium feed-in tariffs were dismantled and replaced with less subsidised panels, it made more economic sense for the household to consume, rather than export, as much of their solar electricity as possible. As a result, more panels are now being installed facing east, to help power the breakfast peak, and west, for the evening.

Anyone who enters the solar PV market will quickly discover fierce competition between different panel and inverter technologies. A solar PV system has two parts the solar panels themselves, and the inverter, which converts the DC created by the panels into AC for your appliances and the grid. Modern solar panels last 10 to 15 years. They will slowly degrade in performance over this time, while good inverters now come with a 10-year warranty. Like most consumer goods, with both panels and inverters you get what you pay for. Panels with a higher output cost more, while more durable inverters cost more too. As the active piece of power electronics, a decent inverter is where consumers should not cut corners. It's also common sense to make sure the company that supplies and installs the equipment will be around over the next 10 years in case something does go wrong.

The two strategic questions for prospective solar PV customers are (1) size and (2) whether a system should be installed that allows for upgrades or new technologies later on. We've noted that the average Australian household consumes around 16 kilowatt hours of electricity per day, but this is only an average. The way households consume electricity is anything but constant. Roughly half is consumed during the day, and half at night. The rates at which different-sized households consume electricity also varies: small flats and apartments tend to use less, while large suburban houses with families, pools and air conditioners tend to consume much more, with big surges in demand when air conditioners are worked hard to keep big houses warm or cool. It is on the roofs of these larger, more energy-hungry suburban homes that most solar PV is installed.

A large residential rooftop solar PV system is typically sized at around 5 kilowatts. This will cover around 30 square metres, much of the available space on most suburban roofs. A well-designed 5-kilowatt system will, on average, generate around 20 kilowatt hours during each day. The actual day-to-day output varies significantly depending on location and season. On longer, clear, mild sunny days a system can generate double this, while on shorter, cloudy days it will produce only 10 to 25 per cent of its rated output, down to a few kilowatt hours of generation. There are also losses: around six per cent of output is lost from the inverter and wiring, five per cent from dirt build-up on the panels and 10 per cent in hot weather (solar panels work better when it's cool).

This means that solar generation and household demand tend to be only partly aligned. On sunny days, solar output will regularly exceed what the household consumes during the day, and on cloudy days it may not produce enough. Regardless of whether a solar PV system has been busy or quiet during the day, it does not work in the dark. As a result of this mismatch, most solar households are constantly using the electricity network, exporting their surplus electricity during the day and importing electricity at night. The occasional talk by rooftop solar owners of going 'off-grid' could not be further from the truth.

A solar PV system always prioritises powering its own house first. Only when all the demand from the house has been supplied does the inverter then dispatch the surplus into the grid. Therefore new solar PV systems deliver the greatest value when they displace retail electricity purchased at around 25–30 cents per kilowatt hour. For these rooftop solar systems, the export value of surplus solar electricity varies, but is based on the wholesale price of electricity (around 7–10 cents per kilowatt hour).

As a result of this, choosing the size of a new solar PV system depends on a range of factors. Larger systems tend to be more lucrative when a household's daytime electricity demand is more constant, because people are home during the day (as with retirees), or if they are going to attach the system to a battery. Smaller systems tend to have a better payback for households that aren't using as much power during the day. While no two households or rooftop solar systems are quite alike, it's typical to get a five-year payback on a well-configured household solar PV system. Which is why there are two million of them on Australian roofs right now.

The big new question for consumers contemplating new solar installations is what to do about home batteries. Rooftop solar systems can link in with batteries to store surplus electricity (rather than send it to the grid) and use it later to offset more retail grid electricity. Household batteries are available now, but are still relatively expensive for the value they provide. If and when they get cheaper, existing solar owners are likely candidates to retrofit a battery to their system.

It is possible to install a battery-ready system. This uses a hybrid inverter that can bolt on a battery some time in the future. These inverters are more complex and more expensive, currently adding more than $1000 to the cost of the installation, which increases the time taken to get payback on the system. The risk here is that this extra cost is wasted if the owner does not end up installing the battery as planned. If home batteries do get cheaper they can always be added later with their own inverter. These are typical of the technology risks consumers face all the time in fast-evolving consumer goods markets, like buying desktop computers in the 1990s or mobile phones in the 2000s.

One issue solar consumers should be aware of, but mostly aren't, is voltage. The grid is supposed to operate at 230 volts (240 volts in Western Australia). In practice, voltage varies slightly across the grid: if a house is near a substation or a street has a lot of solar PV already operating on it, chances are at times the voltage will be higher than 230 volts. When a neighbourhood has lots of solar PV, this means many of these solar houses are pushing surplus electricity onto the grid at the same time. When local demand isn't big enough to use this power, it pushes up local voltages. With more than two million rooftop solar systems in Australia, these new high voltages are becoming a material issue for network providers and, therefore, consumers.

As part of an overhaul of national standards to manage safety and power quality in the grid, since 2016 solar inverters have been limited to operate at a maximum of 255 volts (for what is called 'sustained over-voltage'). If the grid voltage in the street is running higher than this level, the solar power from a household won't be able to get onto the grid, and the inverter will trip off. The prevalence of high voltages was highlighted in recent trials of a virtual power plant (VPP) run by AGL in Adelaide. A VPP is an extension of the smart grid concept. Subscribing solar PV houses across Adelaide were fitted with home batteries and then their generation and dispatch were orchestrated remotely by AGL. The trial was designed to explore how distributed generation could work in the future, by providing both savings to households through more efficient use of their solar generation, and the ability to combine the dispatch of small household batteries across the grid to perform the same frequency services as large-grid batteries.

During the trial AGL discovered around 12 per cent of the VPP systems were consistently switching off because they were trying to push into voltages greater than 255 volts. More than half the systems in the trial recorded voltages above 253 volts. Most of these high-voltage recordings were during the middle of the day when solar output was high and demand was low. In other words, there was so much solar in large parts of the South Australian grid that it was preventing other systems from getting on. In some cases these high-voltage pressures can cause solar inverters to continually switch on and trip off during the day, reducing the amount of electricity they are sending to the grid, value for their owners and the working life of the inverter.

From a prospective solar customer's perspective it would be prudent to get a voltage check (at midday, not 6pm). If the street voltage is consistently more than 253 volts, then a new rooftop solar system may struggle to perform optimally against such high voltage pressures. At these numbers, the network is technically over the

Australian standard, and the network operator is obliged to reduce the pressure below the required standard. This involves modifying local transformers – an additional cost imposed on the network by high penetration of solar PV.

High-voltage management is just one of a number of unforeseen consequences of fast and loose renewable energy policy in Australia. It is solvable but raises deeper issues. Solar PV systems are a consumer good, but they are also part of the electricity grid. To date, solar households have largely been indulged without consideration of what impacts this would have on the grid. This approach is becoming unsustainable. The design of the grid needs to be determined by what will provide electricity to meet reliability standards and emissions reductions at the lowest possible cost. We have planning laws that restrict how landowners can develop their property. Maybe we need planning laws for the grid too. Many network providers would like the ability to limit solar PV installations in parts of the grid where they are over-supplied and voltages are already too high – or at least to allow them only when installed with batteries.

Rooftop solar PV systems have already been subsidised by federal and state governments, and their owners have been under-charged for the fixed cost of using the network. High voltages pose additional costs on all electricity customers because of a lack of planning of distributed generation. Solar PV is a mature technology that will continue to grow in the 21st century. It, and technologies like it, is part of the grid and need to be treated as such. They need to work for the system as a whole, not just their owners. Solar generation needs to be encouraged where it is most valuable, and constrained where it imposes greater costs than benefits. Continued solar under-regulation just accelerates us towards an electricity grid that is out of control; a runaway train powered by a heady mixture of populism, ignorance, self-interest, good intentions and political expedience.

Going 'off-grid'

Rooftop solar PV may be wildly popular, but it is only a partial solution. The sun doesn't always shine when you need electricity. Storing some of a household's surplus solar electricity to be used later during the evening peaks is a logical answer to this problem. A battery smooths out demand by storing some of the surplus electricity for use at times of spikes in demand, reducing the amount of electricity bought from the grid. As welcome as batteries would be, the barrier is price: a 10-kilowatt-hour battery costs more than $10,000. In a perfect world, if a 10-kilowatt-hour battery could recharge and discharge fully every day of the year for 10 years, it could save its owner around $11,000 in total. But the world isn't perfect. Most batteries can't fully discharge and they don't quite reach and hold their rated efficiency. Also, this number doesn't factor in the value of the surplus PV sold back into the grid, and the fact that there will be cloudy days when there isn't enough surplus solar to recharge the battery. As a result, home batteries are still below the payback water line: they cost more than they save.

Still, a house with solar and batteries can be configured to isolate from the grid if there is a local blackout – another advantage. This enables the household to keep using the remaining electricity in the battery until mains power is restored. This extra fit-out costs nearly $1000, but early indications suggest it is a popular option with consumers. It also raises important questions about how a more distributed grid will need to work in the future. In particular, it asks us to consider the difference between being grid-connected and the notion of becoming completely electricity independent – going truly 'off-grid'.

In the thrall of the increased electricity independence that solar PV systems can provide, it's been accompanied by the idea that getting 'off-grid' – a suburban household completely cut off from the grid – is some kind of energy nirvana: the ultimate act of defiance by individuals against big energy companies and governments. But a household with a (large) 5-kilowatt solar PV system and a (current standard) 10-kilowatt-hour battery will not produce and store enough electricity to go it alone. They will still need to buy electricity from the grid. Becoming fully self-sufficient for electricity requires both more generation and more storage to maintain supply at night times, and during extended dark and cloudy periods in winter when the solar PV is producing at a reduced output and during periods of high demand.

If a household wants to go truly off-grid and not incur chronic blackouts and scarcity of electricity for extended periods, it will face the same challenges as the national market, just scaled down. First, it will need enough generation and storage to meet its demand peaks. A 5-kilowatt solar PV plus 10-kilowatt-hour battery combo has no chance of supplying enough electricity for a typical suburban household throughout the year. If a household installed three full days' storage capacity (which is tight) for average consumption of 16 kilowatt hours a day (low for the average household), it would require a 10-kilowatt solar PV system (requiring double the average roof space of a solar PV system) and around 48 kilowatt hours of battery (and the batteries will need to be stored somewhere too). To round down current costs, let's call this $60,000.

Even then, this system doesn't guarantee full supply during periods of sustained cloud or high demand. In these events, off-grid houses will be tempted to run a diesel generator to top up their power. Either that, or they'll have to occasionally ration their electricity use. This system assumes no new major loads – electric cars, for instance. So to qualify for the 'off-grid club' a consumer will need to own a large house with room for a shed, and a spare $60,000. At times an off-grid household will over-produce. Sunny, mild spring days will fill the batteries and spill – nowhere. That zero-emissions electricity will be effectively wasted.

This notion of households opting, at significant cost, to disconnect from the grid also poses existential questions about what the electricity grid is in the 21st century. It is many things: a machine, a market and an essential service. Uniting all these, the provision of electricity is part of the society we live in. A household can be disconnected from the grid at home, but those consumers will still use the grid when they drive down lit streets, go to work and school, shop and eat out. There is a societal element of electricity systems that has always existed, but never been noticed. Until now.

Being connected to and using the grid is not only okay – it's essential. For distributed generators being part of the electricity machine is a co-dependent relationship. Just as solar-and-battery houses still rely on mains power, so the grid will increasingly need to be able to get help from and coordinate these distributed systems at critical times. Being grid-connected enables households with storage to access new revenue streams. There are important mutual benefits.

The societal importance of the grid is most visible during times of peak demand, when every bit of generation is being coordinated to make sure the lights stay on. Not just for houses and air conditioners, but for those neighbourhood services such as traffic lights, hospitals and supermarkets. At these critical times, off-grid consumers won't be contributing anything. Their banks of batteries combined together could be vital. If there is a blackout for whatever reason, this will be their moment of triumph, when all that excessive spending and virtue signalling finally pays off. They can sit at home with their lights on and air conditioner working, while everyone else is in the dark. Ironically, they could have got the same outcome from just buying a diesel generator.

Going off-grid in remote Australia is a necessity. But cutting the wires while living inside the grid should not be permitted. This is antisocial behaviour in the delivery of a social good. At a critical time, when everyone in the grid needs to pull together, some would prefer to cut and run. It also encourages inefficiency: distributed batteries at scale can deliver real benefit for the grid. They will also more handsomely reward their owners, smooth peaks and, if coordinated, can provide valuable ancillary services. At key parts of the grid that are expensive to service, the development of independent micro-grids and even off-grid households may be more efficient and beneficial. This is a planning decision for the networks that provide this service. The grid is not any one company or government. It is society. Like planning for solar PV, we should think carefully about all the rules for all participants.

Electric vehicles

Like home batteries, electric vehicles offer a tantalising glimpse of what the future may hold in personal transport. Electric cars are proving to be better than conventional combustion-engine vehicles in a number of aspects: they're safer, more efficient, cleaner, quieter; the weight in their batteries can be distributed lower so they hold the road better and they can be recharged without producing greenhouse gas emissions. As range improves, the biggest constraint on their market penetration is their price. And this is largely because of the cost of their batteries.

Lithium-ion and other similar compounds are lightweight chemical battery technologies originally developed for small portable devices and then adapted for larger ones (cars).

Battery engineers have been improving efficiency, increasing cycle rates and reducing the share of scarcer elements such as cobalt in their production. But they have, as yet, been unable to bring battery costs down far enough to enable electric cars to be priced competitively with combustion-engine vehicles. Most major car manufacturers have concept cars and production-line electric cars ready for this price moment. But still we wait.

The drive-away price for Tesla's Model S electric sedan starts at $140,000. It remains the safest passenger car ever tested, primarily because of the absence of a large combustion motor in the front of the car. Tesla engineers used the space at the front of the car as a giant crumple zone, without having to worry about diverting the engine out of harm's way in the event of a head-on collision. On release, the more accessible Nissan Leaf (Mk1) sold for AUD$51,000. While a sharper price than the Tesla, it was expensive for what it was. Nissan's similar-sized petrol hatchbacks sold for half that.

If electric cars do cross this price horizon, this will have important consequences for all consumers. Electric cars can be recharged at home. Households in Australian middle and outer suburbs are well suited to accommodate this: most have off-street parking (for the recharger) coupled with a high proportion of two-car households. Anxiety around whether the batteries will have sufficient range has proven to be a non-event: 98 per cent of Australian car trips are well inside the range of even the shorter ranged electric cars. There are now more than five million plug-in electric vehicles worldwide. In particular, Norway, with a population of 5.4 million, has more than 150,000 electric vehicles on its roads. The streets of Oslo are not littered with electric cars that have run out of power.

First-generation electric cars are unidirectional, which means they can only be charged from the grid at designated recharge facilities. Second-generation electric cars will be able to be configured to charge and discharge into the grid as well. Electric cars reinforce the value of a grid. They will be cheaper to run if they can recharge during times of cheap abundant electricity (windy nights, mild sunny afternoons). This value proposition will be accelerated by more dynamic tariffs, creating a new and useful source of demand at times when renewable electricity is spilling over the sides of the glass. They will be similarly discouraged from recharging at times of high demand, except in an emergency – like buying cornflakes from a 7-Eleven.

What's next for consumers?

The transformation taking place across the 21st-century electricity grid will actively include consumers. Millions of households and businesses are already generating some of their own electricity, and thousands are already working with networks and other third parties to have some of their devices controlled to help smooth peak loads. Nearly half of Australian households now have a smart meter, and the sooner the rest get one the better. This will help the grid work better and help renters and home owners alike to manage and reduce their own consumption. They will open the gate for load shifting and dynamic tariffs.

The grid will need to monitor and sometimes control some rooftop solar PV systems and home batteries to help with power quality and, maybe one day, aggregated for bulk power too. The biggest challenge will be how to avoid a two-speed grid, where passive 20th-century customers such as renters, flat owners and the vulnerable continue to be price takers and remain excluded from the technologies that deliver greater efficiency, flexibility and cheaper energy bills. Meanwhile tech-savvy home owners will continue to enjoy the benefits of being active 21st-century prosumers, exploiting the opportunities provided by abundant cheap electricity while becoming mini-sellers during times of peak demand and scarcity.

In all the policy bedlam since 2006, we are only beginning to consider broader moral and societal questions about the roles and responsibilities of a transforming electricity grid. The grid is not some controlling Big Brother. It's part of the physical systems that define how we work together as a society. There is no looming zombie apocalypse that justifies households being allowed to cut themselves off from this shared service. In any case, anyone willing to spend enough money to go off-grid will almost certainly be eschewing the financial benefits of providing services to the grid while only limiting their own access to new technologies and services. The rights of consumers are important, but should not be given precedence over the needs of society. If ever there was a need for governments to show leadership, it is here and now.

10

WHERE DO WE GO FROM HERE?

Right now it's easy to not be super optimistic about the future of the electricity system in Australia. Power bills, climate change, reliability and policy feuding have constantly been on the front pages of major newspapers. Electricity is a headline issue on the political agenda, the subject of strong and widely differing opinions and debate. Australians are starting to learn the names of the oldest coal-fired power stations like they remember the names of the seven dwarves.

Electricity was launched into the political stratosphere by climate change – or, more specifically, by a middle class panicked at the personal impact of severe drought. This anxiety was successfully converted by Kevin Rudd into a political bow wave on climate change action and a landslide electoral victory in 2007. Rudd's backroom politics redesign of energy policy was more potent than he, or most others, thought it would be. His 'Grand Designs' were eventually thinned out by political compromise, until only the most popular elements survived. If Kevin McCloud were to come back to Australia ten years after the original build, he would find a structure unrecognisable from the plans drawn up in 2007. No one remotely planned or anticipated that Australian electricity would end up like this. But nonetheless, here we are.

We will never know what would have happened if the original, carefully planned and designed emissions trading scheme had survived in 2009. Would we have had a renewables target? Would we be where we are today? The idea was to have a market-driven, planned and bankable decarbonisation across the entire sector over decades. Instead we ended up forcing in thousands of megawatts of wind and solar, and we watched as more than half of them squirted into the smallest and least-integrated part of the NEM. Even then, governments only reacted when the system started to break. We have also encouraged millions of Australian households to put solar panels on their homes without any plan about what this would do or how it would work. It was simply allowed to keep going because it was so popular.

As a result of this the grid is in a slow-moving crisis. The intervention and uncertainty have deadened the market signals needed to get critical new investment in generation. Australia's coal-fired power stations are shuffling towards their use-by date, and we have no plan for replacing them. Power prices spike upwards with every closure. The fragility of the grid becomes more acute each summer as reserve margins for generation get tighter and policy-induced blackouts begin. State and federal governments bicker with increasing ferocity about the merits of their rival interventions.

Behind this pantomime, the grid is slowly degrading. Electricity is inadvertently decarbonising in Australia, but not as cheaply, or reliably, as it should. Every town and major city has a rooftop black-ops solar power station that continues to expand yet can be neither monitored nor controlled. As a result, voltages in some suburban streets are chronically high because of unregulated installation of new distributed generation.

These distributed solar power stations are already taking some parts of the grid far into uncharted waters. Perth's isolated grid is predicted to be powered completely from rooftop solar when it reaches minimum-demand events by the middle of the 2020s. In 2018 the AEMO warned a parliamentary inquiry that urgent reforms were needed to manage the uncontrolled growth and uncontrolled generation coming from solar PV. The state's electricity hive mind is trying to work out how the world's most isolated large-scale grid will operate safely and reliably under these conditions. It will probably require the large-scale replication of the King Island micro-grid.

It's obvious with the benefit of hindsight that plenty of mistakes have been made over the past decade. The wonderful part is how much we've learned from them. The opportunity now is what we do with this crazy set of accidental experiments that have defined 21st-century electricity in Australia.

The simplest guide to good electricity policy is to remember the obvious: electricity is not a cause or a political ideology. It's a machine. It's not a contest between renewable energy and conventional, between rooftop solar and power stations or between various rival political 'targets'. The electricity grid is a machine that has been disrupted by climate risk. This has already happened. It is beyond the remit of state or federal governments to decide otherwise. They can only choose if and how they manage this risk so that the machine keeps working as reliably and affordably as possible.

The machine is owned collectively by governments, hundreds of companies and millions of home and building owners. It is both an essential service and a market. Its continued successful operation is a fundamental input to the economy. More than that, it's one of the critical glues that binds our society together.

The pathway to resetting the machine so it can solve for the most efficient outcome, factoring in the new climate constraint, involves three steps. First, admit we have an electricity problem. Second, make a credible plan that incorporates climate risk and stick to it. Third, create the right political and commercial conditions that enable investment to proceed and allow governments to stop trying to design and then build the grid themselves.

The problem

My name is Australia, and I have an electricity problem. I'm running out. This has been going on for more than a decade. Over the past twelve years I have tried six times to get it together, to get a national plan for climate and energy. But every time I backslide. I've tried trading and taxes, targets and obligations. When I hit rock bottom I even tried a thing called Direct Action, whatever that was. None of it worked. And now the blackouts have started …

Governments do two things: they lead or react. Genuine leadership feels more like a memory – civil rights reforms; international engagement; abolishing inequity and injustice; major economic reforms such as deregulation, currency and tax reform; social reforms such as providing access to universal health care and education, or gun control. Genuine leadership seems harder these days, but climate change requires it, and so does energy reform. Australian governments didn't lead on climate change, they reacted to it. We've been reacting ever since.

If this continues, the default future is more of the same: a political electricity grid. That means new power stations and transmission lines and other essential infrastructure will be commissioned, or co-funded or built by any one of seven state and territory governments. Operating independently of one another, they will supply electricity assets that they and the electorate can see and understand: new power stations, renewable theme parks, nation-building pumped hydro, transmission lines and household subsidies. The National Electricity Market (NEM) will retreat to being simply a dispatch mechanism. The big winners will be lobbyists and public relations firms. The backing of a technology or project will depend heavily on how it is perceived. This approach will produce an electricity grid designed by focus groups.

The motivation to react is strong for governments. In response to rising wholesale electricity prices, the Tasmanian Government said they were leaving the NEM. They weren't, of course – it was just a gaudy way of re-announcing continued suppression of the state's wholesale prices. Leaving the NEM would be a truly daft idea for Tasmanians. The state's electricity supply was left in a critical state in 2015 when they were accidentally cut off from the NEM for six months after the undersea cable to Victoria failed.

In 2017 the Queensland Government used its dominant position in generation to artificially suppress wholesale prices, providing price relief for Queensland customers but weakening even further the investment signals for new generation. The Victorian Government signalled its ideological opposition to onshore gas development by banning all conventional and unconventional development, sterilising gas supply that would help bring gas and electricity prices down. Taking this one step further, they then intervened to save manufacturing jobs by cutting a deal to prevent the closure of the state's biggest electricity customer, the Portland aluminium smelter. This deal occurred while the Hazelwood power station was closing. This intervention tightened the electricity market in Victoria at a critical stage and pushed up wholesale electricity prices for the whole state. These specific interventions were just recent highlights of a seemingly never-ending motorcade of populist announcements made without public consultation, such as subsidies for household energy technologies and big renewables targets.

These interventions in the electricity market were made to solve short-term political problems (neutralise anti-gas campaigners, job losses, high power bills). There was little regard for the market or efficiency implications. There was no process of consultation or public discussion. The use of a surprise government policy announcement is generally a dead give-away that it is designed to address a political problem.

It is the motivation for these interventions that underscores the problem. The future design of electricity policy needs to be based on how it most efficiently and reliably delivers energy, including emissions reductions. Electricity doesn't work so well if it's turned into a popularity contest. That's how we got into this mess in the first place. The risk is that governments continue to only back new assets that voters can easily understand, and that *appear* to solve problems. The real value of these will vary, but the choice to back them will be made with at least one eye on the polls.

As a result, an electricity grid designed by politics will be less efficient, more expensive and will divert billions of dollars of government funding from other services. To try to defray the costs, governments will contract some of these energy projects out to the private sector. But they will still need to be underwritten.

Governments will be stuck footing most of the $100-billion-plus rebuild costs.

This will, incidentally, unwind the competition policy reforms put in place in the 1990s. These were a key part of the microeconomic reforms that helped shield Australia from the extremes of economic shocks over the past three decades. The grid will gradually devolve from an integrated machine to a weakly coordinated jumble of pet projects with bits and pieces bolted on, emergency repairs and backup generation. Without an agreed plan, successive changes of government will keep shifting the grid's direction. It will be hard to avoid having Australia's energy devolve into a bloated, lurching and increasingly chaotic system characterised by the pathology of effusive press releases and a never-ending blame storm for inevitable failures on price, emissions and reliability. Australia risks becoming an international case study in how not to decarbonise: the Kodak of electricity.

Of course, we don't elect politicians to design or run complex machines like an electricity grid. That's not their job. It's certainly not their area of expertise. The job of governments is to set strategic direction, like a company board, and get experts to work out the details. In the case of energy, their role has evolved to creating and supporting a market to deliver the service of electricity and gas. A government's role in managing the climate disruption is to set targets on emissions supported by carefully designed, efficient mechanisms to deliver them. None of this is supposed to come near the operation of the machine itself. That's a bit like sitting on a domestic flight approaching Brisbane airport and having Nationals MP George Christensen walk into the air traffic control tower and say, 'Can I have a go?'

The plan

In the words of Led Zeppelin, there's still time to change the road you're on. Plan B returns this design-and-build process to the market, which we built to do this job, under the stewardship of the relevant technical agencies. This has been the intention of the two most recent attempts to defuse the political bomb of climate and energy policy: the Finkel Review, and the National Energy Guarantee (NEG). At the time of writing, the NEG is still in play as a policy option. It's had a curious journey, supported at different stages by both major parties, just not at the same time. At its core, the NEG is the Jedi mind trick of climate and energy policy: it has been designed to be sufficiently effective, yet appear as inoffensive to both sides of politics as possible.

The NEG places the requirement on electricity retailers to demonstrate that they have bought electricity that is backed by enough firm generation in the event of major demand spikes like heat waves, and is also compliant with emissions targets. If there are overs or unders, they can trade with each other, providing the net result still meets these obligations. The threshold for how much firm generation is required is set by the market operator (AEMO) and the emissions targets by the federal government. If there isn't enough firm generation, or enough renewable generation, then the retailers will need to go and build some more, or find someone who will. As a last resort, the AEMO can step in and commission new renewables and/or firm generation to ensure the obligations are met. Usefully, renewable energy projects and the new generation of fast-dispatch gas peakers can both be built quickly, in around a year to 18 months.

There has already been detailed design work on the scheme and there will likely be further technical teething problems around things such as how forecasting demand and supply is calculated, who builds and pays for 'exotic' technology requirements such as synchronous condensers, and the details around what happens if AEMO has to step in. But at least these decisions will be made by technical experts, not politicians. A NEG backed by both major parties reduces risk for private investment. As they have done previously, electricity companies will be leaned on to show their love for the NEG by committing to new generation projects and cutting power bills. If it becomes the new plan, it will be worth it.

The Energy Security Board

An NEG would be the centrepiece of a suite of reforms needed to guide new investment in generation. Its implementation will be necessary, but not sufficient. Over the past decade, electricity and climate change have expanded significantly in the public consciousness. Millions of Australians – politicians, industry leaders, activists, consultants, academics, commentators, celebrity businesspeople and just thousands of ordinary punters – have formed a range of variously informed views on the subject. Government ministers of both colours have extolled the virtues of their most recent energy policy announcements with the confidence that only the cloak of ignorance can provide.

In the modern 24-hour news cycle, governments are relentlessly reacting to media reporting and creating media events to direct the news cycle. This need for constant messaging and strongly worded, simple claims sits poorly with the process of considered design and investment needed to get the best outcome in electricity. Regardless of the merits of this idea, governments will continue to fill the media void out of sheer political necessity unless someone, or something else does. It's like a law of political physics: something has to occupy that political vacuum, or governments will instinctively be drawn back in.

This was one of the rationales behind the Finkel Review's recommendation to create an Energy Security Board (ESB). There is already a cohort of expert regulatory, operational and policy technicians managing the grid. It's just most of us don't see or notice them. Few wish to be regular public communicators or are comfortable intervening in fast-moving political and media debates. An expert may be confident in boffin-to-boffin discussions and arguments but be completely incapable of managing the politics of a 24-hour news cycle.

There are three technical agencies running the grid: a market operator (the Australian Energy Market Operator or AEMO), which handles planning and day-to-day operations; a regulator (the Australian Energy Regulator or AER), which sets the price for monopoly businesses such as transmission and poles and wires, and which regulates behaviour in wholesale and retail markets; and a rule maker and policy adviser (the Australian Energy Market Commission or AEMC). They provide measured, expert advice and reports on their various areas of expertise. While they may form strong opinions about particular interventions or proposals, their views are mostly hidden in technical code. They wrestle over technical and policy challenges in a language that few outsiders understand.

All this is important, but it doesn't fill the political vacuum. Electricity needs a visible, public-facing entity, chaired by a Reserve Bank Governor type of figure, whose primary job is to constantly and sagely inform and reassure Australians that the electricity reform is being managed successfully by experts, while being able to respond quickly to political or technical debates about the operation of the grid.

This is where the Energy Security Board (ESB) could come in. The current ESB is a pop-up board made up of the chairs and CEOs of these three agencies (and two independent board members). It has been designed to implement the Finkel Review's reform process and then dismantle. It needs to be made permanent, and it needs to be redesigned to coordinate the technical horsepower of the electricity A-team and use it to fill this critical role as the go-to agency for comment on electricity issues. Achieving this will be the ESB's (or whatever its re-evolution is called) measure of success. This role of neutral go-to adviser is currently vacant. It is temporarily filled by anyone who looks like an expert: lobby groups, think tanks, activist bloggers, activist academics. This means pretty much anyone with an official-sounding title and a catchy 10-second sound bite message gets a run, regardless of accuracy.

This public-facing role of the ESB should not be limited to media. A critical and relentless task is for the ESB to become a proactive scrutineer of various reports and claims made in the energy debate. Over the past decade, a fantasy electricity system has been created around the real one like scaffolding. This fake grid is constructed from claims made by dozens of reports written with dozens of separate agendas (including self-interest, self-promotion, rent-seeking, selling/defending a project or technology, activism and political distraction). From claims about wind turbines making people sick to Australia having the highest electricity prices in the world (in 2018 the International Energy Agency reported this honour belonged to Germany), and from exaggerated claims about subsidies for renewables to exaggerated claims about subsidies for coal.

Big claims get a lot of traction in this debate because most consumers of information don't know enough to filter fact from fiction. In the land of the blind, the one-eyed man is king. Some of these claims get recycled to support even more outlandish and less accurate positions. Some of these can even end up part of major political speeches and policy commitments. Managing public expectations starts with a trusted, reliable base of information that is credible and neutral.

To support this, the ESB should become the librarian and publisher of all key electricity data in Australia. Currently, this data is located all over the place: it is collected by industry bodies, held by the Clean Energy Regulator, the Australian Bureau of Statistics, and produced in different reports by the AEMO, the AER and the AEMC. Network, generation, clean energy and retail businesses hold quite detailed data, some of it on the changing amounts of electricity used by different sectors of the economy. The network businesses could quite clearly see the decline of the Australian car industry years before it closed, simply because they could track sustained declines in electricity demand. This kind of data provides greater insight into the changing shape of the economy, corrects false and misleading claims, and, in doing so, provides better information to existing and potential new entrants into competitive markets. Information should be an enabler, not a barrier, for potential disruptors.

As well as guiding the public face of reform, correcting misinformation and organising the various data streams, the ESB should also be given oversight of agencies created to support the reform process. In 2011, the Australian Renewable Energy Agency (ARENA) was created, followed by the Clean Energy Finance Corporation (CEFC) in 2012. At the time, their focus was on how to accelerate and mature the development of renewable energy technologies. ARENA was in charge of targeting promising technologies that needed research and development funding to accelerate them out of the lab and into commercial trials. The CEFC was supposed to follow on after this, helping expensive but proven new renewable technologies get down the cost curve by supporting them into the market and finding some economies of scale.

These noble objectives didn't quite pan out as planned. There were plenty of dud new renewable technology ideas, but very few good ones. As solar PV and wind kept thundering down the cost curve, these new renewable technology concepts just fell further and further behind. Realising this, ARENA broadened its focus to providing more detailed information on new supporting technologies and helped to fund trials of a much broader range of activities: demand response, storage, off-grid, even industry policy. The downside to ARENA's work is that it has been increasingly broad and too polite, like an enthusiastic kindergarten teacher: everyone got an elephant stamp. ARENA needs to be a lot more candid in pointing out the adequacies and inadequacies of potential technologies. This means getting reports written by neutrals who are not advocates and devotees of the technologies they are reviewing.

The CEFC was conceived in a political storm at the end of the Gillard government. To survive, it just rolled into a zero-risk financing ball and replicated the role of the commercial banks, funding to return a profit so that a hostile government had no rationale to wind them down. This worked, but it minimised their impact. In any case, there weren't a whole lot of new technologies coming through. The focus for new technology development has shifted radically in the past six years. As it turns out, we won't need the Melbourne Cup field of renewables that was originally conceived. The big challenges now are in fine-tuning power quality with high-penetration intermittent renewables, and working out how to store or generate bulk electricity with zero emissions. The CEFC may be required to move quickly to finance the investments needed. It may need to take more measured risks. That's what it was created to do.

The ESB should also oversee any evolution of market design. This is a technical debate between experts on whether the existing design is capable of signalling and supporting the scale of investment needed. If it cannot, then what is the best way to augment this? What is the most efficient way to finance new, large and fast generators that only work for a few hours a week, or a few weeks a year? There are also new giant pieces of equipment being built – synchronous condensers, resistors and storage. These don't fit neatly in the old model. Who builds them? Where? How much is enough? Who pays for them? As the grid and the economy changes, how do we efficiently replace the stabilising inertia currently supplied by power stations and big electric motors? How do we let someone disrupt all this if they discover a 'much better way'?

The radical changes happening to the grid invite radical solutions. Most large-scale solar PV projects are being located on sites towards the western edge of the network spanning the east-coast of the continent. It's a curious development. The further west of a city a solar panel is located, the deeper into the evening peak its electricity can be generated. Applying this logic, how far west could large-scale solar one day be located? Will it ever be cost effective to run a 1000-kilometre-long transmission line to connect solar in Central Australia to power the evening peaks along the eastern seaboard? This kind of thinking should be welcomed but then expertly scrutinised. There is already a growing cohort of tyre kickers and press-release power station builders (who make grand announcements but never build anything). The ESB needs to be more proactive and more incisive.

Taking control

The new ESB needs to ensure the AEMO can see and control the entire grid when it needs to. In 2017 the AEMO wrote a report effectively requesting greater ability to monitor and control distributed energy technologies. Remote monitors and controllers may need to be progressively fitted to some or all household batteries and solar PV systems, with the ability to retrofit existing systems. Part of this debate will be around who controls them – the retailer, the network, the market operator, or all three? Consumers will be indifferent to the governance options. What is important is that households understand the new terms and conditions before purchase.

Network owners also need to be able to direct new planning rules for the install of new household solar and batteries. This means in some cases constraining new rooftop solar systems where they are already causing over-voltage problems, and be able to encourage more solar and/or batteries where they would augment the grid. This is particularly relevant at the edges of the grid where it is often cheaper to augment supply locally than to build additional poles and wires.

Once these planning limits have been approved, households need to be advised of the new rules. These are the types of possibly unpopular but necessary reforms that governments are currently reluctant to do themselves. Consumers who want to supply the electricity grid in the 21st century need to be part of the planning of the 21st-century grid. It's a shared asset. Their investment needs to be maximised not just for their own bill, but for the system as a whole. We accept speeding limits when we drive and council approval before we build. This isn't any different.

The ESB may also want to require more transparency around the operation of coal-fired generators as they approach their last years of operation. Part of the anxiety around recent closures was about the short notice given: six months for Northern and Hazelwood. In 2018, rules were brought in requiring a minimum of three years' notice of closure. But that misses the point slightly. Coal-fired power stations can still close quickly if they fail environmental or safety standards. Despite the anxiety, the problem with these closures wasn't actually the notice period so much as the remarkable lack of replacement generation. The AEMO will need to establish a more active relationship with twilight coal generators. It will need to ensure new capacity is on line and ready as soon as old capacity is retired. Some flexibility may be required either to squeeze another year or bring forward the closure date.

The end of coal

For the past two years, various ministers in the Turnbull and Morrison governments have talked up large but unspecified commercial support for building a new coal-fired power station in Australia. Labor and the Greens publicly committed to blocking such a proposal. This is just trash talk. A coal-fired Punch and Judy show. Each of the political protagonists is just playing to their respective political bases. In reality, it is extremely unlikely that a new coal-fired power station will ever be built in Australia. This is not an ideological perspective, it's a legal and commercial perspective. Here's why.

On 13 December 2018, the Morrison government called for registrations of interest in a new 'Underwriting New Generation Investments' program. In effect, this was framed as a federal government scheme to underwrite new firm generation: forget the market, Scott Morrison is going to choose. Proponents had six weeks over Christmas to lodge their ideas, with the evolving design of the scheme still to come at the time of publication. It was a political fix, not a technical one.

The only proposal they got for a coal-fired generator was from senior energy entrepreneur Trevor St Baker. St Baker knows his way around the market. Having created the energy company ERM power, he then sold out and cleverly bought the Vales Point power station from the New South Wales Government for $1 million in 2015. The subsequent tight market conditions increased its book value to $730 million. St Barker is no slouch. But nor does he have the capital, or the scale of business, required to get a multi-billion-dollar coal-fired generator through all the planning, protests, politics and construction.

Even if St Barker was to get the go-ahead from the Morrison government, it doesn't have any legal jurisdiction over planning or energy in Australia. Under the Constitution, these have both remained with the states and territories. The federal government only has the ability to legislate nationally against international greenhouse reduction treaties, such as emissions trading and renewable energy targets. A new coal-fired generator would need to receive relevant state or territory government planning approval. No current government has indicated that it would be supportive of such a proposal. The federal government has recently acquired full control of Snowy Hydro, but for the purpose of this exercise, Snowy is just another energy company. It would still require the same planning approval as anyone else.

If a new coal generator was to proceed, then the location would be determined by where the power station would be most valuable and where there was access to suitable and sufficient coal. This isn't as easy as it sounds. A coal-fired power station in North Queensland makes little sense. North Queensland already has around 800 megawatts (about half a large coal-fired power station) of gas generation that has been mothballed. Firm generation is needed much further south. The two main coal-fired power stations to close so far have been Northern (South Australia) and Hazelwood (Victoria). The next to close will be Liddell (New South Wales) in 2022. Then Vales Point (New South Wales), Gladstone (Queensland) and Yallourn (Victoria) are likely to follow some time towards the end of the 2030s, unless forced out earlier.

So Victoria is the epicentre of where new firm generation is needed. But both Victoria and South Australia only have brown-coal reserves, which have higher emissions and more carbon risk than black coal. They also both have state parliaments that are not predetermined to backing new coal generation. The most strategic location, which could utilise black coal and is closest to the main source of new demand, is in New South Wales. But many prospective sites are either owned by companies that do not want to build coal or are constrained by lack of access to coal. It may be possible to build a new coal-fired power station on the banks of Lake Macquarie, assuming its new owners can contract coal supplies and rail access to bring it to the proposed power station.

So what if a suitable place is found? Well, it takes around seven years to build a coal-fired power station. In basic terms, for the project to proceed, St Barker or other proponents are going to need the support of both the federal and New South Wales governments for most of this planning, approval and construction process.

It's likely this timeframe will be delayed as much as possible by sustained interventions from environmental campaigners. This development process would need to endure two state and two federal elections. The last time this 'window' of federal–state pro-coal alignment might have existed was for about 15 months in 2013–14 under the Abbott federal government and Newman government in Queensland. It is difficult to see it re-emerging any time in the immediate future.

The longer it takes to build coal, the less likely that it will ever be built. Coal (and nuclear) are technically incompatible with increased wind and solar PV generation. They are simply too inflexible to adjust to the variations in generation. As renewables increase their generation in the grid, they will be putting increased commercial and operational pressure on the remaining coal generators. Even if firm generation is required, coal would need to be largely quarantined in the market from intermittent generation. This would be more feasible if there was a matching industrial load, like an aluminium smelter, that could contract the generation and cover the risk for decades into the future. There is no aluminium smelter in Australia with anything that remotely resembles that kind of business plan. Instead, until late last year, the biggest investor in aluminium production in Australia, Rio Tinto, had been unsuccessfully trying to sell out of the struggling business since 2015.

What does this mean for coal mining in Australia? Absolutely nothing. Australia produces some of the highest-quality power-station and steel-making black coal in the world. It produces less greenhouse gas when combusted than coal from most other parts of the world. We will need to continue to supply our remaining domestic power stations, and supply export coal to customers, mainly in Korea and Japan, until they close their coal power stations. Shutting down Australian coal mining won't make things any better. It will just make Australia poorer, and global emissions higher.

What about emissions?

The objective of Australia's energy policy should be for the sector to meet national emissions commitments while maintaining reliability at the lowest possible cost. That's it. This transformation is about emissions, not renewables. Renewables generation is now a mature technology. A sign of the maturing of the debate is that installing new renewables generation is increasingly seen as unremarkable. There will be times when wind and solar provide most of the electricity required, and there will be times when they will produce nothing. Installing renewables generation should not be portrayed as some sort of contest. It is simply a feature of the evolving electricity system. Renewables targets or commitments should be abolished. In a successfully repaired electricity market, they will be redundant.

Almost all employment relating to renewable energy occurs at the time of installation. Once operating, these technologies are highly automated and reliable. They have lower permanent workforces than coal-fired power stations because there isn't much to do. In any case, we don't want more jobs in renewable energy. We want fewer jobs. The fewer people we have to pay to install and produce electricity, the cheaper it becomes. Cheap energy is one of the most powerful economic forces on the planet. It reaches far beyond lower power bills, creating opportunities for new industries and processes.

A more useful focus for governments would be a strategy to improve the energy performance of the 400,000 public housing dwellings across the country. Many of the most vulnerable people in our society live in public housing. A coordinated strategy to reduce their electricity consumption through a combination of greater efficiency, flexibility and distributed generation is, on any measure, a more appropriate use of government subsidies than throwing more money at middle-class home owners.

Australia: the first mover

Australia has been dealt a peculiar hand on electricity. Our abundant, high-grade coal underwrote our electricity system in the 20th century. The unwinding of coal in the 21st century poses new challenges: on the upside we have lots of gas, it's windy and it's sunny. Australia sits above the roaring forties winds that rip around the Southern Ocean, and it's hot and sunny across much of the inland areas and the west of the continent. On the downside, we are dry and remote. We have a finite and relatively small amount of hydroelectricity available to firm renewables, and we can't plug into anyone nearby when things get tight. We are also a major uranium exporter, but the closest we have come to nuclear power was building a car park near a beach. Through political accident and falling technology costs, Australia finds itself careering towards hosting the world's biggest intermittent renewable energy grid. Early next decade, we are approaching the very real prospect of running a major grid like Perth's on pure renewable generation during minimum-demand events. Planning is already underway to sort out the supporting suite of grid-stabilising technologies needed in the next five years. This isn't a controlled trial. This show is going live. Soon.

No other major grid in the world is close to this yet. California sources around 19 per cent of its electricity from wind and solar, but it imports around 30 per cent of its electricity from Canada and other US states. Texas has a more isolated main grid, but only sources around seven per cent of its electricity from wind, peaking at up to 40 per cent of demand. In Europe, perhaps the closest grid to Australia's is Ireland's, which sits at the edge of the European grids and now sources 25 per cent of its electricity from wind energy. It has two interconnectors to Scotland and Wales and is exploring another 500-kilometre-long interconnector directly into the nuclear reactors of France (Basslink between Tasmania and Victoria is around 370 kilometres). The problem facing the Irish electricity planners is growing uncertainty about whether there will be enough spare capacity on the French end of the transmission lines at times when they need to import extra electricity. It's a bit like South Australia.

There is the technical challenge of ensuring there are enough synchronous condensers, giant capacitors and batteries to help maintain power quality during these all-intermittent-renewable events. Far beyond these technical challenges, the bigger impact is on the economy. What does this mean for Australian manufacturing? For its big remaining electricity customers in metals processing? Instead of having inane arguments about whether we should build a new coal-fired generator or promising renewable energy jobs that don't exist, maybe we should be thinking about what Australia will look like as an economy by mid-century. It will be an economy framed by a vastly different electricity system – potentially with periods of abundant, cheap electricity followed by periods of scarcity and high prices.

This requires a change in mindset. Last century we built industrial processes to match the constant supply of cheap electricity produced by large thermal power stations. Australia needs a radical rethink in the development of industry policy: how best to manage existing industries during this transition, and which industrial processes might be adapted to fit the new generation system in the 21st century. When European farm machinery manufacturers were asked to build wind turbines for California in the 1980s, this sparked the creation of a new, world-leading industry. If Australia is going to go first running these new types of electricity grids, how could we commercialise it? These are the jobs we want to be creating in the 21st century.

This line of thinking has already begun. German aluminium smelter TRIMET Aluminium SE has already built a prototype smelter that it claims works like a giant battery. The electrolysis furnaces that use large amounts of electricity to make aluminium are clad with heat exchangers, which store and release heat energy. Other inventions include the development of idling aluminium smelter cells that can load surplus electricity and discharge at peak times, effectively using the aluminium to store the energy. In Australia, Alcoa is working with the University of Adelaide to look at using the heat from solar thermal concentrating technologies to make aluminium.

Is gas our next best option?

The NEM is going to need around 15,000 megawatts of new firm-equivalent capacity over the next 15 to 20 years. This is a bulk-power requirement. It is well beyond the capacity of the most advanced batteries currently available, and is more than seven times more power than can be supplied by the Snowy 2.0 pumped hydro proposal – if it proceeds. At this stage most of this firm generation will need to be supplied by gas generators, which means Australian governments are soon going to face a more serious reality check about the state of the gas market, and some of the constraints they have imposed on gas supply.

Regional Australia is increasingly finding itself part of the 21st-century energy industry. Key regions are hosting increasing numbers of large-scale solar PV farms, wind farms and unconventional gas projects. Sometimes these communities have reacted to this rapid change with concern. As a reflection of the lack of planning, no one really thought to properly consult these communities about the changes heading their way. Resulting campaigns against wind and gas have been so successful because they have filled this vacuum. Their advice has too often been the only advice.

Plans to build a few dozen 40-storey-high turbines along a ridge line are fine until you have to live near them. Wind and gas developers have been reluctant to publicly admit they sometimes mismanaged early developments, scorching the earth for those who followed them. Regional Australia is hosting this transformation. They deserve a better seat at the table and a clearer plan about what is coming next.

We also need to stop demonising gas. Gas is a renewables enabler. We will need it to support the high-renewables grid we are building right now. The role of gas will diminish when technologies to provide cleaner bulk power evolve and displace it. The faster this can happen, the faster deeper emissions targets can be reached. Until then, we will need to build a lot more highly flexible firm generators, assisted by grid-scale batteries for power quality. This requires a reset on access to eastern Australia's abundant conventional and unconventional gas fields. Either that or we will need to ship gas in from overseas while we ship it out through Gladstone. It's a messy look. We can do better.

Finally

Australia is on the brink of a major change to the way it powers its economy. This will bring a shift in the direction of the economy itself. The current transformation of electricity is much, much bigger than electricity bill shock or special deals for rooftop solar. It will reshape the way Australia works, what it does in the world and how we are perceived on the world stage.

Some Australians may like to fantasise that we are seen as a nation of rugged Crocodile Dundees. In the real world, Australians have a reputation for being relaxed and down to earth. Australian senior diplomats (and Canadians) are regularly recruited to chair difficult meetings at international negotiations because they somehow bring their national traits to the room: they are respected neutrals who want an outcome, don't indulge in puffery and are comfortable forcing the room to make a decision. No one hates us. We're seen as a collection of mines with glorious beaches around them. Traditionally we're technology followers, not leaders. We wait and move with the pack. Some of our most notable inventions reflect our lifestyle: the Hills hoist, the lawn mower.

Suddenly, Australia finds itself riding in a renewable energy breakaway from the peloton. Australia's legacy of high-emissions generators has made it an easy target of derision for some activists. As much a surprise to ourselves as anyone else, the peculiar climate and energy chaos of the past decade finds us suddenly making a material break from the developed world on climate and energy. Australia's electricity is still a paradox: generation is still dominated by ageing coal power stations, while wind and solar continue to scale up at pace. Cities such as Adelaide and Perth now find themselves unwittingly at the cutting edge of high-renewables integration. This wasn't in anyone's plan. We just followed someone's wheel and, well, here we are.

What do we do now? We literally can't go back. So we can only go forwards. The brains trust of Australia's electricity boffins seem unfussed by the technical challenge; they're more concerned with trying to make things work while the rules keep changing and governments keep changing their minds. Most of the technology is available now. We can automatically shift loads so that most customers don't notice we are doing it. We can exploit our comparative advantages of having big wind, sunshine and gas and rebuild a reliable, affordable electricity grid that uses these – and an economy that exploits these too. It is likely to be different to the economy that exploited cheap, reliable coal-fired electricity in the 20th century. That's okay. Not everything has to be the same.

Australia has a truly unique opportunity right now. We can reorganise our electricity grid and our economy. We can move first and be world leaders in industries that can adapt and thrive in intermittent generation grids. This is likely to integrate into decarbonising transport systems, and then the need for techniques to reduce emissions in agriculture. However we choose to solve this, the real mark of success will be when people stop talking about electricity. When it becomes invisible again. And that will be because, after we've finally stopped arguing and got on with a sensible plan to decarbonise the grid, it just works.

Acknowledgements

This book is the synthesis of more than a decade of meetings with, listening to conference presentations by, reading reports written by and interviewing or asking questions of hundreds of experts across the debate on climate science and the energy industry in Australia and overseas. There are simply too many people to remember to name individually. In almost all of these conversations I seemed to be the one asking dumb questions, which they answered with remarkable patience and kindness for which I am grateful.

I'd like to specifically thank Rob Jackson, Nick Leys, Andrew Leunig and Duncan Mackinnon for reading through earlier drafts of this book and politely correcting technical errors or, even worse, incorrect use of apostrophes. I'd also like to specifically thank Ben Skinner for providing technical advice on electrical engineering over the past decade.

My thanks also go to Paul Simshauser AM, Tim Nelson, Ray Massie, Peter Cosier, Jessie Foran, Kieran Donoghue, Tim Duignan, Erwin Jackson, Ivor Frischnecht, Cameron Parrotte, Andrew Dillon, Colin Wain, Brad Page, Andrew Richards, Sarah McNamara and arl Kitchen for specific information, clarification and general bouncing ideas off. And thanks to Ian MacFarlane and Martin Ferguson for your time as well.

Finally I would like to thank Valeria Lema for putting up with me with great patience and love as I wrangled these words together.

Selected references

Acil Allen Consulting (2017), Energy Consumption Benchmarks Electricity and Gas for Residential Customers, Melbourne, AER.

Acil Allen Consulting (2018), Peak Demand and Energy Forecasts for the South West Interconnected System Western Australia, Melbourne, AEMO.

AECOM (2014), Australia's Off-Grid Clean Energy Market Research Paper, Canberra, ARENA.

AGL Energy Ltd, ElectraNet Pty Ltd, WorleyParsons Services Pty Ltd, (2015) Energy Storage for Commercial Renewable Integration South Australia, ARENA.

AGL Energy Ltd (2017), Virtual Power Plant in South Australia, Stage 1 Milestone Report, Sydney, AGL.

AGL Energy Ltd (2018), Virtual Power Plant in South Australia, Stage 2 Public Report, Sydney, AGL.

Arup (2014), Smart Grid, Smart City: Shaping Australia's Energy Future, Sydney, Ausgrid.

Australian Bureau of Statistics (2017), Introduction of the 17th series Australian Consumer Price Index, 6470.0.55.001, Canberra.

Australian Competition and Consumer Commission, (2018), Restoring Electricity Affordability and Australia's Competitive Advantage, Canberra.

Australian Energy Council (2018), Submission to the South Australian Energy Transformation Project Assessment Draft Report, 31 August 2018, https://www.energycouncil.com.au/submissions.

Australian Energy Council, Energy Networks Australia (2018), Electricity Gas Australia 2018.

Australian Energy Market Commission (2018), Reliability Frameworks Review, Sydney, AEMC.

Australian Energy Market Commission (2018), Review of the Regulatory Frameworks for Stand-Alone Power Systems, Sydney, AEMC.

Australian Energy Market Operator (2013), 100 Per Cent Renewables Study – Modelling Outcomes, Melbourne, AEMO.

Australian Energy Market Operator (2016), National Electricity Forecasting Report, Melbourne, AEMO.

Australian Energy Market Operator (2016), Market Modelling Methodology and Input Assumptions, Melbourne, AEMO.

Australian Energy Market Operator (2016), South Australian Renewable Energy Report, Melbourne, AEMO.

Australian Energy Market Operator (2017), Black system South Australia 28 September 2016, Melbourne, AEMO.

Australian Energy Market Operator (2017), Electricity Forecasting Insights, Melbourne, AEMO.

Australian Energy Market Operator (2017), South Australian Electricity Report, Melbourne, AEMO.

Australian Energy Market Operator (2017), Visibility of Distributed Energy Resources, Melbourne, AEMO.

Australian Energy Market Operator (2018), 2018 Electricity Statement of Opportunities, Melbourne, AEMO.

Australian Energy Market Operator (2018), 2018 Gas Statement of Opportunities Melbourne, AEMO.

Australian Energy Market Operator (2018), Integrated System Plan for The National Electricity Market, Melbourne, AEMO.

Australian Energy Market Operator (2018), Quarterly Energy Dynamics Q3 2018, Melbourne, AEMO.

Australian Energy Market Operator (2018), Summer 2018-19 Readiness Plan, Melbourne, AEMO.

Australian Energy Regulator (2018), AER Electricity Wholesale Performance Monitoring Hazelwood Advice, Melbourne, AER.

Australian Energy Regulator (2018), State of the Energy Market 2018, Melbourne, AER.

Australian Government (2013), Alice Solar City 2008-2013, Canberra, Australian Government.

Australian Government Treasury (2013), Strong Growth, Low Pollution – Modelling a Carbon Price, Canberra, Treasury.

Australian Government Treasury (2017), Analysis of Wage Growth, Canberra, Treasury.

Bailey, I (2007), Neoliberalism, Climate Governance and the Scalar Politics of EU Emissions Trading, The Royal Geographical Society, 39(4), p. 431.

Bailey, I (2009), Market Environmentalism, New Environmental Policy Instruments, and Climate Policy in the United Kingdom and Germany, Annals of the Association of American Geographers, 97(3), p. 530.

Bakan, J (2005), The Corporation, The Pathological Pursuit of Profit and Power, New York, Free Press.

Baxter, R (2007), Energy Storage. Tulsa, PenWell.

Brady, F (1999), A Dictionary on electricity, Contribution on Australia, International Conference on Large High Voltage Electrical Systems (CIGRE).

Britt, A; Summerfield, D; Senior, A; Kay, P; Huston, D; Hitchman, A; Hughes, A; Champion, D; Simpson, R; Sexton, M and Schofield, A (2017), Australia's Identified Mineral Resources 2017, Geoscience Australia, Canberra.

Butler, M (2017), Climate Wars, Melbourne, Melbourne University Press.

CAT Projects (2015), Investigating the Impact of Solar Variability on Grid Stability, ARENA.

Chaney, M (2006), BCA President's Address to BCA Annual Dinner, Sydney, Business Council of Australia.

Clean Energy Council (2018), Clean Energy Australia Report 2018, Melbourne, Clean Energy Council.

Clean Energy Regulator (2018), http://www.cleanenergyregulator.gov.au/RET, Canberra.

Chesworth, J (1960), The electric tramways of Hobart, Sydney, Australian Electric Traction Association.

COAG Energy Council (2018), National Energy Guarantee COAG Energy Council Decision Paper.

Coleman, T (2003) The Impact of Climate Change on Insurance Against Catastrophes, Institute of Actuaries Australia 2003 Biennial Convention.

Commonwealth of Australia (2002), Towards a Truly National and Efficient Energy Market (Parer Review), Canberra, Commonwealth of Australia.

Commonwealth of Australia (2008), Carbon Pollution Reduction Scheme Green Paper, Canberra, Commonwealth of Australia.

Commonwealth of Australia (2017), Independent Review into the Future Security of the National Electricity Market: Blueprint for the Future, Commonwealth of Australia.

Commonwealth of Australia (2018), National Energy Guarantee Final Detailed Design – Commonwealth Elements, Commonwealth of Australia.

Cooper, I (1993), Hobart tramways: a centenary commemoration, Sydney, Transit Australia Publishing.

Department of Energy and Electric Power Research Institute (2013), Electricity Storage Handbook in collaboration with NECRA, Albuquerque, SANDIA.

Department of the Environment and Energy (2018), Australian Energy Statistics, Canberra.

Donoghue, K and Australian Energy Council (2017), Investment in Australia's Electricity Generation Sector to 2030, Melbourne, Newgrange Consulting.

Edwards, C (1969). Brown Power: a Jubilee History of the State Electricity Commission of Victoria, Melbourne, State Electricity Commission of Victoria.

EirGrid Group (2017), All-Island Generation Capacity Statement 2017-2026, Dublin, Eirgrid Group.

ElectraNet (2018), SA Energy Transformation RIT-T Project Assessment Draft Report, Adelaide, ElectraNet.

Elmegaard, B and Wiebke, B (2011), Efficiency of Compressed Air Energy Storage, The 24th International Conference on Efficiency, Cost, Optimization, Simulation and Environmental Impact of Energy Systems.

Energy Networks Australia (2017), Electricity Network Transformation Roadmap: Final Report, Melbourne, ENA.

Energy Queensland (2018), Demand Management Plan 2018-2019, Brisbane, Energy Queensland.

Energy Supply Association of Australia (2015), Australia's Electricity Archipelago: The Challenges of High Renewable Generation in Small Island Grids, Melbourne, ESAA.

Energy Supply Association of Australia (2016), Electricity Gas Australia 2016, Melbourne, ESAA.

Essig, M (2005), Edison and the Electric Chair, Penguin.

Flannery, T (2005), The Weather Makers: The History & Future Impact of Climate Change, Melbourne, Text Publishing.

Gerardi, W and Galanis, P (2017), Report to the Independent Review into the Future Security of the National Electricity Market, Emissions Mitigation Policies and Security of Electricity Supply, Melbourne, Jacobs Group.

Gold, R (2019), PG&E: The First Climate-Change Bankruptcy, Probably Not the Last, New York, Wall Street Journal.

Graham, P et al (2018), GenCost 2018, Updated Projections of Electricity Generation Technology Costs, CSIRO.

Green, M (2000), Power to the people. Sydney, UNSW Press.

Grubb, M; Azar, C and Persson, U (2005), Allowance Allocation in the European Emissions Trading System: A Commentary, Climate Policy, 5(1) p.127-136,.

Hancock, M (2011), Alice springs, A Case Study of Increasing Levels of PV Penetration in an Electricity Supply System, Sydney, University of NSW.

Harrison, K et al (2010), Global Commons, Domestic Decisions: The Comparative Politics of Climate Change, Cambridge MA, MIT Press.

Hesse, H; Schimpe, M; Kucevic, D and Jossen, A (2017), Lithium-Ion Battery Storage for the Grid—A Review of Stationary Battery Storage System Design Tailored for Applications in Modern Power Grids. Energies, 10(12), p. 2107.

House of Commons Library (1993), Carbon Taxes and Global Warming Research Paper 93/106, London, House of Commons Library.

Hyslop, P (2018), Snowy 2.0 – Is the Reward Worth the Risk?, Renew Economy, https://reneweconomy.com.au/Author/Paul-Hyslop.

Intergovernmental Panel on Climate Change (2014), AR5 Climate Change 2014: Mitigation of Climate Change, Contribution of Working Group III to the Fifth Assessment Report of the Intergovernmental Panel on Climate Change, Cambridge, Cambridge University Press.

International Energy Agency (2018), World Energy Outlook 2018, Paris, International Energy Agency.

International Energy Agency (2018), World Energy Prices 2018: An Overview, Paris, International Energy Agency.

International Renewable Energy Agency (2014), Wave Energy Technology Brief, Abu Dhabi, IRENA.

International Transport Forum (2010), Transport Greenhouse Gas Emissions: Country Data, 2010, OECD.

Jacobs Group (2016), Retail electricity price history and projected trends,

Melbourne, AEMO.

King Island Renewable Energy Integration Project (2019), http://www.kingislandrenewableenergy.com.au/.

Kings, K (2016). A Short History of the North Melbourne Tramways and Lighting Company Limited, Melbourne, Nunawading Tramway Publications.

Koh, R and Li, A (2018), Australia Utilities, Australia Clean Energy, Morgan Stanley Research.

Junreuther, H and Michel-Kerjan, E (2007), Climate Change, Insurability of Large-Scale Disasters, and the Emerging Liability Challenge, University of Pennsylvania Law Review, 155 p. 1795.

Lewis, M and Curien, I (2008), Carbon Emissions It Takes CO2 to Contango, London, Deutsche Bank Global Markets Research.

Lovegrove, K; James, G; Leitch, D; Milczarek, A; Ngo, A; Rutovitz, J; Watt, M, and Wyder, J (2018), Comparison of Dispatchable Renewable Electricity Options, Canberra, ARENA.

Lowy Institute, (2018), Lowy Institute Poll 2018, Understanding Australian Attitudes to the World, Sydney.

Lynn, P (2013), Electricity from sunlight. Hoboken, Wiley.

Macintosh, A and Wilkinson, D (2010), The Australian Government's solar PV rebate program – An evaluation of its cost-effectiveness and fairness, Policy Brief No. 21, ANU Centre for Climate Law and Policy.

Mercer, D (2018), Rooftop Solar Poses Blackout Threat to WA's Main Power Grid, The West Australian, May 21 2018.

Meyer, H (1971), A History of Electricity and Magnetism, Cambridge, MIT Press.

Morrison, P (1934), How Electricity Came to Melbourne, The Argus, October 16 1934, p. 37.

National Renewable Energy Laboratory (2018), US Solar Photovoltaic system Cost Benchmark: Q1 2018, Denver, NREL.

Parliament of Australia, (2009), Coalition Dissenting Report on the CPRS Changes with Additional Comments by Senator Joyce, Leader of the Nationals in the Senate on Behalf of the National Party, Canberra, Senate Standing Committee on Economics.

Pera, M (1992), The Ambiguous Frog: The Galvani-Volta Controversy on Animal Electricity, Princeton, Princeton University Press.

Perlin, J (2002), From Space to Earth: the Story of Solar Electricity, Cambridge, Harvard University Press.

Pierce, M and Mandelbaum, J (2009), Early Electricity Supply in Melbourne, 3rd Australasian Engineering Heritage Conference, Dunedin.

Productivity Commission (1999), Microeconomic Reforms and Australian Productivity: Exploring the Links, Melbourne, Productivity Commission.

Productivity Commission (1995), Study into the Australian Gas Industry and Markets, Canberra, Productivity Commission.

State Electricity Commission of Victoria (1989), Annual Report, Melbourne.

Ragleb, M (2014), Solar Thermal Power and Energy Storage Historical Perspective, https://www.solarthermalworld.org/sites/default/files/story/2015-04-18/solar_thermal_power_and_energy_storage_historical_perspective.pdf.

Royden, A (2002), U.S. Climate Change Policy Under President Clinton: A Look Back, Golden Gate University Law Review, 32(4), p. 415-478.

Sandiford, M; Forcey, T; Pears, A and McConnell, D (2015), Five Years of Declining Annual Consumption of Grid-Supplied Electricity in Eastern Australia: Causes and Consequences. The Electricity Journal, 28(7), p. 96-117.

Schmidt, O; Hawkes, A; Gambhir, A; and Staffell, I (2017), The Future Cost of Electrical Energy Storage Based on Experience Rates, Nature Energy 2.

Simshauser, P and Doan, T (2009), Emission Trading, Wealth Transfers and the Wounded Bull Scenario in Power Generation, The Australian Economic Review, 42(1), p. 64-83.

Simshauser, P; Nelson, T and Doan, T (2011), The Boomerang Paradox, Part 1: How a Nation's Wealth is Creating Fuel Poverty, The Electricity Journal, 24(1), p. 72.

Simshauser, P and Nelson, T (2014), The Consequences of Retail Electricity Price Rises: Rethinking Customer Hardship, The Australian Economic Review, 47(1), p. 13-43.

Simshauser, P, (2018), On Intermittent Renewable Generation and the Stability of Australia's National Electricity Market, Energy Economics, 72(C), p. 1-19.

Speck, S (2008), The Design of Carbon and Broad-Based Energy Taxes in European Countries, Vermont Journal of Environmental Law, 10(1), p.31.

SunShift Pty Ltd (2017), Renewable energy in the Mining Sector White Paper, Canberra, ARENA.

Wilkenfeld, G and Spearitt, P (2004), Electrifying Sydney: 100 years of Energy Australia, Sydney, Energy Australia.

US Energy Information Administration (2018), Levelized Cost and Levelized Avoided Cost of New Generation Resources in the Annual Energy Outlook 2018, Washington, US EIA.

Utilities Commission of the Northern Territory (2010), Power System Review 2008-09, Darwin, Utilities Commission.

Utilities Commission of the Northern Territory (2011), Power System Review 2009-10, Darwin, Utilities Commission.

Utilities Commission of the Northern Territory (2012), Power System Review 2010-11, Darwin, Utilities Commission.

Utilities Commission of the Northern Territory (2013), Power System Review 2011-12, Darwin, Utilities Commission.

Utilities Commission of the Northern Territory (2014), Power System Review 2012-13, Darwin, Utilities Commission.

Utilities Commission of the Northern Territory (2015), Power System Review 2013-14, Darwin, Utilities Commission.

Utilities Commission of the Northern Territory (2016), Power System Review 2014-15, Darwin, Utilities Commission.

Utilities Commission of the Northern Territory (2017), Power System Review 2015-16, Darwin, Utilities Commission.

Utilities Commission of the Northern Territory (2018), Power System Review 2016-17, Darwin, Utilities Commission.

Vestergaard, J et al (2004), Industry Formation and State Intervention: The Case of the Wind Turbine Industry in Denmark and the United States, Academy of International Business Conference Proceedings.

Wyborn, D; de Graaf, L; Davidson, S and Hann, S (2005), Development of Australia's First Hot Fractured Rock (HFR) Underground Heat Exchanger, Cooper Basin, South Australia, World Geothermal Congress 2005, Antalya.

Australia, 3, 81-84, 101, 110, 139, 154, 186

Tasmanian Government, 37, 233
Telecom Australia, 121
Telstra, 38
Territory Generation, 193
Tesla, 21, 105-106, 151, 153-155, 185, 226
Three Mile Island, 159
Turnbull, Malcolm (and Government), 57-58, 86-87, 103, 140, 142, 151, 244
TXU, 40

Vail, Alfred, 25
Vales Point, 106, 245-246
Vanguard satellite, 119
Vestas, 125
Victorian Electric Light Company, 29
Victorian Energy Efficiency Target, 204
Volta, Alessandro, 23
voltage, 13, 27-29, 32, 110-112, 142, 176, 178, 184, 187, 219-221, 231

Weatherill, Jay, 78, 101-102, 106
Whitlam, Gough, 120
wind energy, 3, 10, 12-18, 72-82, 85-86, 90-93, 95-100, 105, 112-113, 115-117, 119, 124-130, 132-134, 138, 140-143, 146-147, 150-151, 153, 158, 163, 167, 181, 184-189, 194, 196-198, 230, 239, 241, 247-254
Wong, Penny, 55, 73

Yallourn power station, 34, 246

Zema, Matt, 80

Program (KIREIP), 128, 183-190, 193-195, 199, 231

Kyoto Protocol, 9, 11, 49-55

Latham, Mark, 49, 53, 72

Latrobe Valley, 6, 33, 35, 78, 83

Liddell power station, 88, 101, 106, 142, 147, 246

Liquefied Natural Gas, 145

Lithium ion, 153, 155, 163, 197, 226

MacFarlane, Ian, 73

Mandatory Renewable Energy Target, 53, 56, 128, 167-168

Marshall, Steven, 105, 156

McLachlan, Gillon, 127

Monash, Sir John, 34

Montreal Protocol, 46

Morrison, Scott, 87, 104, 244-245

Murdoch, Rupert, 127

Murrays Beach car park, 159

Musk, Elon, 105, 153-154

NASA, 46

National Electricity Market (NEM), 4, 39-40, 79, 83, 89-94, 96, 99-102, 103, 108-111, 113-114, 129, 156, 183, 199, 230, 233, 251

National Energy Guarantee (NEG), 86-87, 103, 108, 141, 236-237

National Hydrogen Plan, 156

Newman, Maurice, 127

Nissan Leaf, 226

nuclear energy, 8-10, 16, 48-49, 54, 77, 122, 126, 141-142, 159-162, 164, 247, 249

O'Farrell Government, 177

Obama, Barack, 57-58

Oceanlinx, 133-134

Origin Energy, 40, 136, 145

Pacific Gas and Electricity, 62

Parer, Warwick, 53

Paris Agreement, 86-87

Parkinson, Martin, 55

Pearce, John, 86

Pearson, Gerald, 119

Pelican Point power station, 79, 143

Playford, Tom, 34, 99

Reliability and Emergency Reserve Trader (RERT), 113

Renewable Energy Target (RET), 14, 56, 72-73, 75-76, 78, 97, 101, 137, 140, 177

Rio Tinto, 196-197, 247

Rudd, Kevin (and Government), 11, 49, 55-58, 68, 71-74, 108, 173, 177, 229

Santos, 144-145

Severn Barrage, 133

Shelley, Mary, 23

Shorten, Bill, 156

Siemens, 125

Simshauser, Paul, 8

Smith, Dick, 127

Snowy Hydro, 37, 40, 71, 245

Solar Cities, 190, 191

solar hot water, 73-75, 117, 147, 168, 178, 191, 205-206

Solar Power Corporation, 120

Solar PV, 10, 12-15, 17, 20, 27, 45, 53, 72-78, 81, 85, 90-93, 95-98, 102, 105, 112-113, 115-124, 126, 129-130, 132, 134, 138, 140, 142-143, 146-148, 150, 154-155, 163, 165-179, 187-199, 202-203, 205-207, 209, 215-223, 224-225, 228, 230-231, 241-243, 248-249, 252-253

solar thermal, 87, 101, 105, 147-149, 163, 189, 199, 251

SolarReserve, 149

South West Interconnected System (SWIS), 4, 195

Soylent Green, the movie, 45

St Baker, Trevor, 245

State Electricity Commission, 34

Suntech, 170

system black in South

Energy Developments Limited, 135
Energy Information Administration (US), 138
ENGIE, 83
Eraring Power Station, 17
Ergon Energy, 194
European, 10, 45, 48, 50, 52, 72, 122, 250-251
Exxon, 120

Faraday, Michael, 21, 24, 27,
Faulkner, John, 48
Finkel, Alan, 85-86, 103, 108, 138, 140, 142, 236, 238-239
Flannery, Tim, 63
Franklin, Benjamin, 21
frequency, 7, 13, 61, 100, 110-111, 142, 184, 186-188, 192-193, 220
Frydenberg, Josh, 84-86, 101-102
Fukushima, nuclear accident, 9-10, 159-160
Fusion energy, 162

Galvani, Luigi, 21, 23
gas (as fuel and generation) 3, 12, 14-15, 25, 30, 35, 37-38, 54, 69, 74, 77, 79-84, 88, 90-92, 94, 97, 101-104, 106, 115, 122-123, 126, 141, 143-147, 154-155, 157-159, 163-164, 183, 191, 193-194, 201-205, 233-236, 246, 249, 251-253

Geodynamics, 136-137
geothermal, 73-74, 135-138, 199
Gillard, Julia (and Government), 59, 63, 71, 137, 199, 241
global financial crisis, 10-11, 57, 72, 170-171
gold plating of networks, 111, 172
Gordon Power Station, 36
Gore, Al, 10, 48, 54, 64
Grand Ethiopian Renaissance Dam, 132
Greenhouse Challenge, 48
Greenhouse Gas Abatement Scheme, 204
Greenpeace, 159
Grove-White, Gerry, 136

Hansen, James, 46-47, 64
Hawke, Bob, 37
Hazelwood power station, 83, 85, 103, 111, 113, 139-140, 201, 234, 244, 246
heat pumps, 74, 178, 205-206
Heywood Interconnector, 79
High Efficiency Low Emissions (HELE), 15, 158
Hill, Robert, 53
Hilmer, Fred, 39
Hinkley Point C nuclear power station, 161
Hornsdale wind farm, 151, 153-154
Howard, John (and government), 3, 10-11, 43, 49, 51-56, 57, 68, 72, 128, 167-169, 173, 175, 190
Hydro Electric Commission, 34, 36
Hydro Tasmania, 71, 128, 151, 184-185, 187-190
hydro, 4, 6, 8, 16, 28, 31, 34, 37, 40, 48-49, 67, 71, 77, 87, 91, 113, 115, 128, 130-132, 134, 141, 144, 150-151, 172, 233, 251
hydrogen, 155-156, 163

inertia, 188, 242
Infigen Energy, 106
interconnectors, 78, 80-82, 94, 101, 105-107, 156, 250
Intergovernmental Panel on Climate Change (IPCC), 47
International Atomic Energy Agency, 161, 239
International Energy Agency (IEA), 2, 239
International Power, 40

Jones, Alan, 127

Keating, Paul, 39, 48, 174
Kennett, Jeff, 40
King Island Renewable Energy Integration

268

Index

Abbott, Tony, 58-59, 71, 86-87, 140, 247

Adelaide Electric Supply Company, 34

AGL, 26, 30, 40, 88, 101-103, 106, 142, 220

Alice Springs, 121, 183, 190-194, 198

Alinta Energy, 79

Allam Cycle, 158

aluminium, 6, 38, 75, 91, 112, 141, 204, 234, 247, 251

Ampère, Marie, 21, 24

Australian Democrats, 53, 167

Australian Energy Market Commission (AEMC), 40, 86, 106-107, 109-110, 212, 238, 240

Australian Energy Market Operator (AEMO), 40, 79-80, 92-93, 107, 109, 112-113, 199, 212, 231, 236-238, 240, 243-244

Australian Energy Regulator (AER), 40, 212, 238, 240

Australian Greenhouse Office, 53

Bakan, Joel, 59

Bali, (UNFCCC COP 13), 49, 55, 57

baseload power, 77, 94, 99, 136, 139, 141-142, 149, 158, 161

battery, 22-27, 92, 101, 105-106, 119, 141, 151-155, 178-179, 184-185, 187-189, 193, 197-198, 208, 218-226, 228, 251

Beazley, Kim, 53, 55

Bell Laboratories, 118

Bell Solar Battery, 119

BHP, 197

bioenergy, 134-135, 141, 167

carbon capture and storage (CCS), 157-158

Carbon Pollution Reduction Scheme (CPRS), 57-58

carbon tax, 12, 14, 48-49, 59, 70-72

Carnegie Wave Energy, 133

Carter, Jimmy, 72

Chaney, Michael, 54-55

Chapin, Daryl, 119

Chernobyl, nuclear accident, 159-160

China Light and Power (CLP), 40

Clean Energy Target, 85-86, 140

climate change, 7-8, 10-11, 20, 41, 43-44, 46-49, 51-55, 57, 59-64, 72-73, 88, 90, 114, 122, 140, 152, 157, 160, 164, 166, 168, 172, 175, 190, 202, 229, 232, 237

climate unease, 62-63

coal, 3, 6-12, 14-15, 17, 25, 30, 33-38, 47-48, 50, 54-55, 67-71, 74, 76-77, 80, 83-88, 90-91, 93-94, 97-101, 104, 107, 113, 115, 122-124, 126, 131, 140-143, 145-147, 152, 156-158, 160-161, 163-164, 172, 175, 182, 187, 197, 201, 206, 209, 229-230, 239, 244-250, 253-254

coalition, 52, 55, 57-58, 85-86, 142, 175

Commonwealth Bank, 39

Copenhagen (UNFCCC COP15), 11, 57-58, 73

CSIRO, 138, 161

Direct Action, 59, 232

Duck Reach power station, 31, 131

dunkelflaute, 130

Edison, Thomas, 21, 26, 29

Einstein, Albert, 118

ElectraNet, 106

Electricity Trust of South Australia (ETSA), 34, 128

267